Sir George Greenhill

Differential and Integral Calculus

Sir George Greenhill

Differential and Integral Calculus

ISBN/EAN: 9783337811242

Printed in Europe, USA, Canada, Australia, Japan

Cover: Foto ©berggeist007 / pixelio.de

More available books at **www.hansebooks.com**

From the Author.

DIFFERENTIAL AND INTEGRAL CALCULUS.

DIFFERENTIAL AND INTEGRAL

CALCULUS,

WITH APPLICATIONS.

BY

ALFRED GEORGE GREENHILL, M.A.,

PROFESSOR OF MATHEMATICS TO THE SENIOR CLASS OF ARTILLERY OFFICERS, WOOLWICH ;
EXAMINER IN MATHEMATICS IN THE UNIVERSITY OF LONDON.

𝕷ondon :

MACMILLAN AND CO.

1886.

CONTENTS.

CONTENTS.

ERRATA.

Page 1, line 5 from bottom,—*for* "(f$x+h$) *read* f($x+h$).

Page 106, line 9 from bottom—

for "$n\int(\sin x)^n dx = -(\sin x)^{n-2}\cos x + (n-1)\int(\sin x)^{n-1}dx$".

read $n\int(\sin x)^n dx = -(\sin x)^{n-1}\cos x + (n-1)\int(\sin x)^{n-2}dx$.

Page 128, Ex. 25, *read*—

$$(Ax^3 + 3Bx^2 + 3Cx + D)(Ay^3 + 3By^2 + 3Cy + D)(Az^3 + 3Bz^2 + 3Cz + D)$$
$$= \{Axyz + B(yz + zx + xy) + C(x+y+z) + D\}^3;$$

and $\frac{1}{3}\int(Ax^3 + 3Bx^2 + 3Cx + D)^{-1}dx + \ldots\ldots$

PREFACE.

THE present Treatise is intended as an introduction to the study of the Differential and Integral Calculus, but will be found to contain what is necessary to know in order to pass on to the subjects which presume a knowledge of the Calculus.

I have endeavoured to make this book suitable not only for the mathematical student, but also for men like engineers and electricians who require the subject for practical applications, to whom even a slight knowledge of the notation and methods of the Calculus is becoming more and more indispensable.

Hitherto in this country the influence of Newton, although the inventor of Fluxions, has been employed to delay the study of this subject, and make a knowledge of it the privilege of a select few ; my object in writing this treatise has been mainly to present the subject in as simple a manner as possible, in order to encourage a larger number of students to cultivate it.

In order, however, to keep the size of the book within reasonable limits, it is assumed that the reader has already acquired a knowledge of the elements of the subject of Algebra, Trigonometry, and Co-ordinate

Geometry, as given, for instance, in the treatises of Hall and Knight, J. B. Lock and C. Smith; accordingly I have at once proceeded to explain the operation and application of Differentiation with as little preliminary explanation as possible.

I have followed the recent American treatises of Rice and Woolsey Johnson, Byerly and J. H. Taylor, on this subject in introducing the notion of Time as an independent variable, and the associated ideas of velocity and acceleration, in order to afford illustrations of the use of the Calculus; this is after all only a return to the Method of Fluxions as invented by Newton, and carried out by Maclaurin and other writers in this country, until supplanted by the notation of the Differential Coefficients of the foreign mathematicians.

The Doctrine of Fluxions is a useful and rigorous method of presenting the elementary ideas of the flow of varying quantities, and is employed in the treatises of Rice and Woolsey Johnson under the name of the Method of Rates; but the notation for a fluxion, for instance $\dot{x}$ the fluxion of x, though easily written is difficult to print, and has the inconvenience of not indicating the independent variable, so that the notation of Leibnitz, $\dfrac{dx}{dt}$ instead of $\dot{x}$, is now used almost universally in printed books; and this notation it is now proposed to print in the form dx/dt, in order to economize space.

The chief novelties in the present work consist, first, in carrying on the subjects of the Differential and of the Integral Calculus together, instead of, as is usual, completing the Differential before passing on to the

Integral Calculus ; secondly, in the use of the hyperbolic functions in conjunction with the ordinary circular trigonometrical functions, in order to preserve an exact analogy, which is not apparent when only the exponential and logarithmic functions are employed.

The notation of sinh, cosh, tanh, etc., to denote the hyperbolic sine, cosine, tangent, etc., has been employed, in accordance with what appears to be now the most universal custom.

I have ventured also, in order to preserve symmetry, to introduce the inverse hyperbolic functions, and, following Byerly, to denote them by sinh⁻¹, cosh⁻¹, tanh⁻¹, etc., by analogy with sin⁻¹, cos⁻¹, tan⁻¹, etc. ; this idea will be found indicated in Bertrand's Integral Calculus, Chapter I., but apparently has not been used, in consequence of the cumbrousness of the notation there employed, namely, sect. sin hyp., sect. cos hyp., sect. tang hyp., etc., instead of the above.

By the use of the direct and inverse hyperbolic functions in conjunction with the direct and inverse circular functions, the Calculus is in my opinion considerably simplified, and the student is led on more naturally and readily to the consideration of the elliptic and other functions. The consideration of these functions is however beyond the scope of the present treatise.

In order to exhibit more clearly the analogy and symmetry between the circular and hyperbolic functions, I have made a digression in Chapter I. on the formulæ of the addition equation (as it may be called by analogy with elliptic functions) of ordinary trigonometry, showing how the formulæ may all be deduced from a single figure, with the corresponding relations of the hyperbolic functions.

Numerous collections of examples will be found throughout the book, introduced at each point to illustrate what has immediately gone before.

The order of arrangement will be found in some respects different to what is customary; for instance, the idea of tracing simple curves from their equations has been introduced into the first chapter, so far as is necessary for the ordinary applications of the Integral Calculus to finding the areas, etc., of these curves; the general theory of curves being resumed in the last chapter. Maxima and Minima also have been investigated without the aid of Taylor's Theorem.

Change of the Independent Variable has only been touched upon where necessary; the general theory of Change of the Independent Variable, as well as Lagrange's and Laplace's Theorems, and the Elimination of Constants and Functions, have been omitted as beyond the scope of an elementary treatise.

I have to thank Mr. A. G. Hadcock, Inspector of Ordnance Machinery, Royal Artillery, for drawing the diagrams, and also for revising the proof sheets and preparing the index.

WOOLWICH,
December, 1885.

LIST OF AUTHORS AND THEIR WORKS CONSULTED IN THE PREPARATION OF THIS BOOK.

NEWTON.—*Principia Mathematica Philosophiæ Naturalis*. Cambridge, 1713.

HAYES.—*Fluxions*. London, 1704.

SIMPSON.—*Fluxions*. London, 1737.

MACLAURIN.—*Fluxions*. Edinburgh, 1742; London, 1801.

WARING.—*Meditationes Analyticae*. Cambridge, 1776.

MEIER HIRSCH.—*Integraltafeln*. Berlin, 1810.

DEALTRY.—*Fluxions*. Cambridge, 1812.

LACROIX.—*Differential and Integral Calculus*. Cambridge, 1816.

PEACOCK.—*Differential and Integral Calculus*. Cambridge, 1820.

HIND.—*Differential Calculus*. Cambridge, 1832.

GREGORY.—*Examples on Differential and Integral Calculus*. Cambridge, 1840.

HEMMING.—*Differential and Integral Calculus*. Cambridge, 1852.

WOOLHOUSE.—*Differential Calculus*. London, 1854.

HALL, T. G.—*Differential and Integral Calculus*. London, 1863.

TODHUNTER.—*Differential Calculus*. Cambridge, 1864.

TODHUNTER.—*Integral Calculus*. Cambridge, 1868.

FRENET.—*Recueil d'Exercises sur le Calcul Infinitésimal*. Paris, 1866.

BERTRAND.—*Calcul Différential et Intégral*. Paris, 1870.

WOLSTENHOLME.—*Differential and Integral Calculus*. London, 1874.

WILLIAMSON.—*Differential Calculus*. London, 1877.

WILLIAMSON.—*Integral Calculus*. London, 1884.

RICE AND WOOLSEY JOHNSON.—*Differential Calculus*. New York, 1879. *Integral Calculus*. London, 1883.

BYERLY.—*Integral Calculus*. Boston, 1881.

BYERLY.—*Differential Calculus*. Boston, 1882.

TAYLOR, J. M.—*Differential and Integral Calculus*. Boston, 1885.

SOHNCKE, L. A.—*Sammlung von Aufgaben aus der Differential und Integral Rechnung*. Halle, 1885.

CLIFFORD, W. K.—*Elements of Dynamic*. London, 1878.

CLIFFORD, W. K.—*Common Sense of the Exact Sciences*. London, 1885.

CHAPTER I.

DIFFERENTIATION.

1. *Definition of a Differential Coefficient.*

If $f(x)$ denotes any function of a variable quantity x, and if $f(x+h)$ denotes the same function of $x+h$ when x receives a small increment h, then the limiting value of

$$\frac{f(x+h)-f(x)}{h}$$

when h is indefinitely diminished is called the *differential coefficient* of $f(x)$ with respect to x, and is denoted by $\frac{df(x)}{dx}$ or $f'(x)$. This may be conveniently expressed as

$$\frac{df(x)}{dx} = \operatorname{lt}\frac{f(x+h)-f(x)}{h},$$

(lt) being the abbreviation employed to denote the limiting value as h is indefinitely diminished and ultimately becomes zero.

Since $(fx+h) - f(x)$ is the increment of $f(x)$ corresponding to the increment h of x, therefore $\operatorname{lt}\dfrac{f(x+h)-f(x)}{h}$ is the ultimate ratio of the corresponding increments of $f(x)$ and x denoted by $df(x)$ and dx, and $\dfrac{df(x)}{dx}$ measures the rate of increase of $f(x)$.

A

2. *Definition of a Function.*

One quantity denoted by y or f(x) is said to be a *function* of another quantity denoted by x when the value of y or f(x) depends on the value of x.

The notation fx instead of f(x) will be used henceforth when the argument consists of a single term like x.

Thus x^2, x^3, x^4, x^n, sin x, cos x, tan x, cot x, sec x, cosec x, vers x, $\sin^{-1}x$, $\cos^{-1}x$, $\tan^{-1}x$, $\cot^{-1}x$, $\sec^{-1}x$, $\operatorname{cosec}^{-1}x$, $\operatorname{vers}^{-1}x$, a^x, log x, sinh x, cosh x, tanh x, $\sinh^{-1}x$, $\cosh^{-1}x$, $\tanh^{-1}x$, etc., are simple functions of x, which we shall require hereafter.

We shall proceed to differentiate them, that is to find their differential coefficients.

3. *Differential Coefficients of the Simple Functions.*

It follows from the definition that

$$\frac{dx}{dx} = \operatorname{lt}\frac{x+h-x}{h}$$

$$= \operatorname{lt}\frac{h}{h} = 1.$$

$$\frac{dx^2}{dx} = \operatorname{lt}\frac{(x+h)^2 - x^2}{h}$$

$$= \operatorname{lt}\frac{2xh + h^2}{h}$$

$$= \operatorname{lt}(2x + h) = 2x.$$

$$\frac{dx^3}{dx} = \operatorname{lt}\frac{(x+h)^3 - x^3}{h}$$

$$= \operatorname{lt}\frac{3x^2h + 3xh^2 + h^3}{h}$$

$$= \operatorname{lt}(3x^2 + 3xh + h^2) = 3x^2.$$

$$\frac{dx^4}{dx} = \operatorname{lt}\frac{(x+h)^4 - x^4}{h}$$

$$= \operatorname{lt}\frac{4x^3h + 6x^2h^2 + 4xh^3 + h^4}{h}$$

$$= \operatorname{lt}(4x^3 + 6x^2h + 4xh^2 + h^3) = 4x^3.$$

And generally

$$\frac{dx^n}{dx} = \mathrm{lt}\frac{(x+h)^n - x^n}{h}$$

(expanding by the Binomial Theorem)

$$= \mathrm{lt}\ \frac{nx^{n-1}h + \dfrac{n(n-1)}{1.2}x^{n-2}h^2 + \dots}{h}$$

$$= \mathrm{lt}\left\{ nx^{n-1} + \frac{n(n-1)}{1.2}x^{n-2}h + \dots \right\} = nx^{n-1}.$$

Here n may denote any number, positive or negative.

4. Again

$$\frac{d\sin x}{dx} = \mathrm{lt}\frac{\sin(x+h) - \sin x}{h}$$

$$= \mathrm{lt}\frac{2\cos(x + \frac{1}{2}h)\sin \frac{1}{2}h}{h}$$

$$= \mathrm{lt}\cos(x + \tfrac{1}{2}h)\frac{\sin \frac{1}{2}h}{\frac{1}{2}h} = \cos x.$$

For

$$\mathrm{lt}\cos(x + \tfrac{1}{2}h) = \cos x,$$

and

$$\mathrm{lt}\frac{\sin \frac{1}{2}h}{\frac{1}{2}h} = 1,$$

if x and therefore h is expressed in circular measure.

5. *Definition.* The *circular measure* of an angle AOP (fig. 1) is the ratio of the circular arc AP to the radius OP.

Denoting the ratio of the circumference to the diameter of a circle by π, then the c.m. (circular measure) of a right angle is $\frac{1}{2}\pi$, of the angle of an equilateral triangle is $\frac{1}{3}\pi$, and so on.

Draw PM the perpendicular from P on OA, and let the tangent to the circular arc at P meet OA produced in T; then according to the definitions of Trigonometry, if θ denotes the c.m. of the angle AOP,

Fig. I

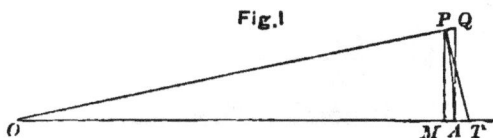

$$\frac{MP}{OP}=\sin\theta,\ \frac{OM}{OP}=\cos\theta,$$

$$\frac{AP}{OP}=\theta,\qquad \frac{OT}{OP}=\sec\theta,$$

$$\frac{PT}{OP}=\tan\theta,\ \frac{AM}{OA}=\text{vers}\,\theta.$$

Now MP, AP and PT are in ascending order of magnitude, and

$$\frac{MP}{PT}=\cos\theta;\ \text{therefore when}\ \theta=0,\ \frac{MP}{PT}=1.$$

Therefore also, when $\theta=0$,

$$\frac{MP}{AP}=1,\ \text{or}\ \frac{\sin\theta}{\theta}=1\ ;$$

and

$$\frac{PT}{AP}=1,\ \text{or}\ \frac{\tan\theta}{\theta}=1.$$

Generally, when $\theta=0$,

$$\frac{\sin m\theta}{n\theta}=\frac{m}{n}\ \frac{\sin m\theta}{m\theta}=\frac{m}{n},\ \frac{\tan m\theta}{n\theta}=\frac{m}{n}\ ;$$

$$\left(\frac{\sin m\theta}{n\theta}\right)^{p}=\left(\frac{m}{n}\right)^{p},\ \left(\frac{\tan m\theta}{n\theta}\right)^{p}=\left(\frac{m}{n}\right)^{p}.$$

Denoting $\sin\theta$ or $\tan\theta$ by h, then when $h=0$,

$$\frac{\sin^{-1}h}{h}=1,\ \frac{\tan^{-1}h}{h}=1,\ \frac{\sin^{-1}mh}{nh}=\frac{m}{n},\ \frac{\tan^{-1}mh}{nh}=\frac{m}{n}.$$

6. Again

$$\frac{d \cos x}{dx} = \text{lt} \frac{\cos(x+h) - \cos x}{h}$$

$$= \text{lt} \frac{-2 \sin(x + \frac{1}{2}h)\sin \frac{1}{2}h}{h}$$

$$= - \text{lt} \sin(x + \frac{1}{2}h)\frac{\sin \frac{1}{2}h}{\frac{1}{2}h} = - \sin x.$$

$$\frac{d \tan x}{dx} = \text{lt} \frac{\tan(x+h) - \tan x}{h}$$

$$= \text{lt} \frac{\sin h}{h}\sec(x+h)\sec x = \sec^2 x.$$

$$\frac{d \cot x}{dx} = \text{lt} \frac{\cot(x+h) - \cot x}{h}$$

$$= - \text{lt} \frac{\sin h}{h}\text{cosec}(x+h)\text{cosec } x = - \text{cosec}^2 x.$$

$$\frac{d \sec x}{dx} = \text{lt} \frac{\sec(x+h) - \sec x}{h}$$

$$= \text{lt} \frac{\cos x - \cos(x+h)}{h \cos x \cos(x+h)}$$

$$= \text{lt} \frac{2 \sin(x + \frac{1}{2}h)\sin \frac{1}{2}h}{h \cos x \cos(x+h)}$$

$$= \text{lt} \frac{\sin(x + \frac{1}{2}h)}{\cos x \cos(x+h)} \frac{\sin \frac{1}{2}h}{\frac{1}{2}h} = \frac{\sin x}{\cos^2 x} = \sec x \tan x,$$

and similarly

$$\frac{d \text{ cosec } x}{dx} = - \text{cosec } x \cot x.$$

$$\frac{d \text{ vers } x}{dx} = \sin x.$$

Examples.—1. Determine from the definition the d.c. (differential coefficient) with respect to x of

$$\sqrt{x}, \; \frac{1}{\sqrt{x}}, \; \frac{1}{x}, \; \frac{1}{x^2}, \; \frac{1}{x^3}, \; \frac{1}{x^m}, \; (x+a)^n, \; \binom{x}{n}^n, \; \frac{mx+n}{px+q}, \; \sin mx, \; \cos\frac{x}{a},$$

$\tan(mx+n)$, $\tan x^2$, $\sin x^n$, $x \tan x$.

2. Obtain geometrically from figure 2 the d.c.'s of the circular or trigonometrical functions.

3. Differentiate with respect to x^m: (i.) x^n, (ii.) $\sin x$, (iii.) $\tan x^n$.

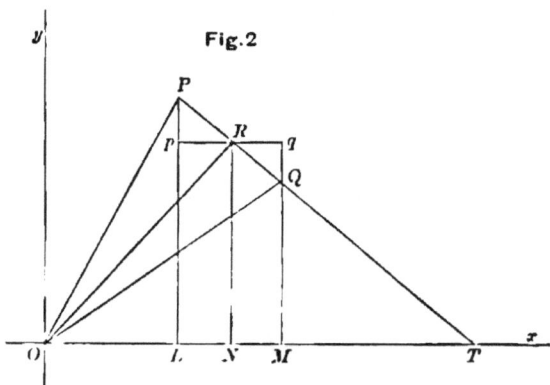

Fig.2

7. The trigonometrical formulæ required in the preceding differentiations can be established as follows from figure 2:—

Bisect the angle POQ by OR, and draw $PRQT$ perpendicular to OR, and complete the figure.

Then, if the angle xOR is denoted by A, and ROP or ROQ by B:

$$\sin(A+B) = \frac{LP}{OP} = \frac{NR}{OP} + \frac{pP}{OP}$$

$$=\frac{NR}{OR}\frac{OR}{OP}+\frac{pP}{RP}\frac{RP}{OP}$$

$$=\sin A \cos B + \cos A \sin B ; \dots\dots\dots\dots(1),$$

since the angle $pPR = NOR = A$.

And

$$\sin(A-B)=\frac{MQ}{OQ}=\frac{NR}{OQ}-\frac{Qq}{OQ}$$

$$=\frac{NR}{OR}\frac{OR}{OQ}-\frac{Qq}{RQ}\frac{RQ}{OQ}$$

$$=\sin A \cos B - \cos A \sin B \dots\dots\dots(2).$$

Adding and subtracting (1) and (2),

$$\sin A \cos B = \tfrac{1}{2}\sin(A+B)+\tfrac{1}{2}\sin(A-B) \dots\dots(3),$$

$$\cos A \sin B = \tfrac{1}{2}\sin(A+B)-\tfrac{1}{2}\sin(A-B)\dots\dots(4).$$

Again,

$$\cos(A+B)=\frac{OL}{OP}=\frac{ON}{OP}-\frac{pR}{OP}$$

$$=\frac{ON}{OR}\frac{OR}{OP}-\frac{pR}{RP}\frac{RP}{OP}$$

$$=\cos A \cos B - \sin A \sin B \dots\dots\dots(5).$$

And

$$\cos(A-B)=\frac{OM}{OQ}=\frac{ON}{OQ}+\frac{Rq}{OQ}$$

$$=\frac{ON}{OR}\frac{OR}{OQ}+\frac{Rq}{RQ}\frac{RQ}{OQ}$$

$$=\cos A \cos B + \sin A \sin B \dots\dots\dots(6).$$

Adding and subtracting (5) and (6),

$$\cos A \cos B = \quad \tfrac{1}{2}\cos(A+B)+\tfrac{1}{2}\cos(A-B) \dots\dots(7),$$

$$\sin A \sin B = -\tfrac{1}{2}\cos(A+B)+\tfrac{1}{2}\cos(A-B)\dots\dots(8).$$

Again,

$$\tan(A+B)=\frac{LP}{OL}=\frac{NR+pP}{ON-pR}$$

$$= \frac{\dfrac{NR}{ON} + \dfrac{pP}{ON}}{1 - \dfrac{pR}{ON}} = \frac{\dfrac{NR}{ON} + \dfrac{RP}{OR}}{1 - \dfrac{pR}{pP} \dfrac{RP}{OR}}$$

(because in the similar triangles pPR and ONR,

$$\frac{pP}{ON} = \frac{RP}{OR}\Big)$$

$$= \frac{\tan A + \tan B}{1 - \tan A \tan B} \dotfill (9).$$

And

$$\tan(A - B) = \frac{MQ}{OM} = \frac{NR - Qq}{ON + Rq}$$

$$= \frac{\dfrac{NR}{ON} - \dfrac{Qq}{ON}}{1 + \dfrac{Rq}{ON}} = \frac{\dfrac{NR}{ON} - \dfrac{RQ}{OR}}{1 + \dfrac{Rq}{Qq} \dfrac{RQ}{OR}}$$

$$= \frac{\tan A - \tan B}{1 + \tan A \tan B} \dotfill (10).$$

Denoting the angle xOP by C, and xOQ by D, so that

$$A + B = C, \ A - B = D,$$

then $\qquad\qquad xOR = A = \tfrac{1}{2}(C + D),$

and $\qquad\qquad ROP = ROQ = B = \tfrac{1}{2}(C - D).$

Then

$$\sin C + \sin D = \frac{LP}{OP} + \frac{MQ}{OQ}$$

$$= 2\frac{NR}{OP} = 2\frac{NR}{OR} \frac{OR}{OP}$$

$$= 2 \sin \tfrac{1}{2}(C + D)\cos \tfrac{1}{2}(C - D) \dotfill (11).$$

$$\sin C - \sin D = \frac{LP}{OP} - \frac{MQ}{OQ}$$

$$= 2\frac{pP}{OP} = 2\frac{pP}{RP} \frac{RP}{OP}$$

$$= 2 \cos \tfrac{1}{2}(C + D)\sin \tfrac{1}{2}(C - D) \dotfill (12).$$

$$\cos C + \cos D = \frac{OL}{OP} + \frac{OM}{OQ}$$

$$= 2\frac{ON}{OP} = 2\frac{ON}{OR}\frac{OR}{OP}$$

$$= 2\cos\tfrac{1}{2}(C+D)\cos\tfrac{1}{2}(C-D)\dots\dots\dots(13).$$

$$\cos C - \cos D = \frac{OL}{OP} - \frac{OM}{OQ}$$

$$= -2\frac{pR}{OP} = -2\frac{pR}{RP}\frac{RP}{OP}$$

$$= -2\sin\tfrac{1}{2}(C+D)\sin\tfrac{1}{2}(C-D)\dots\dots\dots(14).$$

Equations (11), (12), (13), (14) are equivalent to (3), (4), (7), (8) writing C for $A+B$ and D for $A-B$.

Produce PQ to meet Ox in T; then

$$\tan A + \tan B = \frac{RT}{OR} + \frac{RQ}{OR}$$

$$= \frac{PT}{OR} = \frac{LP}{OP}\frac{PT}{LP}\frac{OP}{OR}$$

$$= \sin(A+B)\sec A \sec B \dots\dots\dots\dots(15).$$

$$\tan A - \tan B = \frac{RT}{OR} - \frac{RQ}{OR}$$

$$= \frac{QT}{OR} = \frac{MQ}{OQ}\frac{QT}{MQ}\frac{OQ}{OR}$$

$$= \sin(A-B)\sec A \sec B \dots\dots\dots\dots(16).$$

These formulæ have been proved from a figure in which all the angles are acute and contained in the first quadrant; but the formulæ are seen to be universally true for all magnitudes of the angles, and the proof is letter for letter the same, if we carefully preserve the order of the letters in order to represent the direction of the lines in the figure, as is necessary when lines are used to represent velocities, forces, etc.

8. Putting $A=0$, $\frac{1}{2}\pi$, π, 2π leads to the relations:

$\sin(-B)=-\sin B$, $\cos(-B)=\cos B$, $\tan(-B)=-\tan B$.

$\sin(\frac{1}{2}\pi-B)=\cos B$, $\cos(\frac{1}{2}\pi-B)=\sin B$, $\tan(\frac{1}{2}\pi-B)=\cot B$.

$\sin(\frac{1}{2}\pi+B)=\cos B$, $\cos(\frac{1}{2}\pi+B)=-\sin B$,
$\qquad\tan(\frac{1}{2}\pi+B)=-\cot B$.

$\sin(\pi-B)=\sin B$, $\cos(\pi-B)=-\cos B$,
$\qquad\tan(\pi-B)=-\tan B$.

$\sin(\pi+B)=-\sin B$, $\cos(\pi+B)=-\cos B$,
$\qquad\tan(\pi+B)=\tan B$.

$\sin(2\pi-B)=-\sin B$, $\cos(2\pi-B)=\cos B$,
$\qquad\tan(2\pi-B)=-\tan B$.

$\sin(2\pi+B)=\sin B$, $\cos(2\pi+B)=\cos B$,
$\qquad\tan(2\pi+B)=\tan B$.

Thus 2π is the period of the sine and cosine, but π of the tangent and cotangent.

Also

$$\sin\{n\pi+(-1)^n B\}=\sin B,$$
$$\cos(2n\pi\pm B)=\cos B,$$
$$\tan(n\pi+B)=\tan B,$$

where n is any integer, positive or negative, so that the system of angles

(i.) $n\pi+(-1)^n B$ have the same sine or cosecant;

(ii.) $2n\pm B$ have the same cosine or secant or versed sine ;

(iii.) $n\pi+B$ have the same tangent or cotangent.

9. *Geometrical Interpretation of Differential Coefficients.*

Employing in figure 3 the co-ordinates x and y of a point P referred to axes Ox and Oy at right angles, where $OM=x$, $MP=y$, and x is measured positive to the right, negative to the left, y is measured positive upwards, negative downwards, then the equation

$$y = fx$$

represents some curve CPQ, the assemblage of points

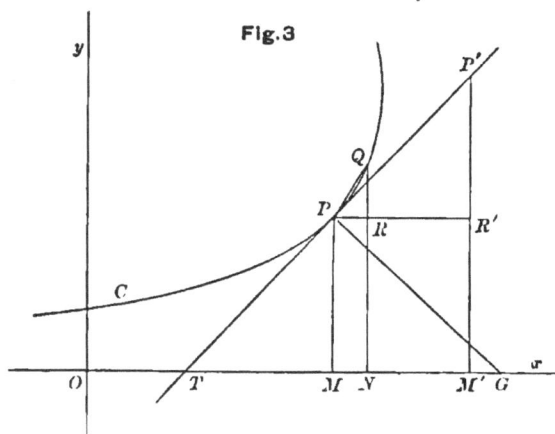

Fig.3

whose co-ordinates satisfy this equation; so that, if $OM = x$, then $MP = fx$.

If $\qquad MN = h$, then $ON = x + h$,

and $\qquad NQ = f(x+h), \; RQ = f(x+h) - fx$;

so that $\qquad \dfrac{f(x+h) - fx}{h} = \dfrac{RQ}{PR} = \tan RPQ$.

Now if the angle xTP which the tangent at P makes with the axis Ox is denoted by ψ, then since the direction of the tangent TP is the ultimate direction of the chord PQ when the point Q has approached indefinitely near to P, therefore

$$\tan\psi = \text{lt} \tan RPQ$$

$$= \text{lt}\frac{RQ}{PR}$$

$$= \text{lt}\frac{f(x+h) - fx}{h}$$

$$= \frac{dfx}{dx} \text{ or } f'x \text{ or } \frac{dy}{dx},$$

since $\qquad y = fx.$

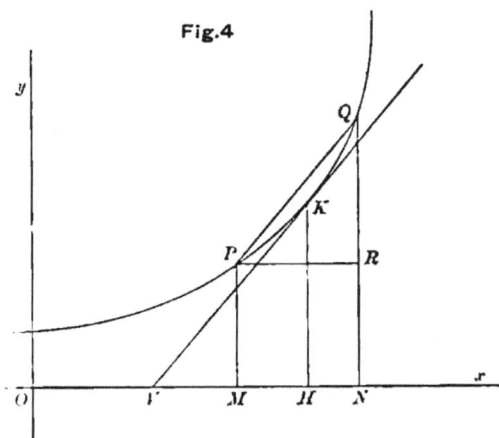

Fig.4

10. Supposing h again, as at first, to denote a finite increment of x, and that fx is a continuous function of x, so that in the curve $y=fx$ the tangent at some point K between P and Q (figure 4) is parallel to the chord PQ; then

$$\frac{f(x+h)-fx}{h} = \tan QPR = \tan KVx,$$

$$= f'(x+\theta h),$$

where $x+\theta h=OH$, the abscissa of K; and $\theta = \frac{MH}{MN}$, a proper fraction, some unknown function of x and h.

Therefore $\qquad f(x+h)=fx + hf'(x+\theta h),$

a theorem required subsequently in Taylor's Theorem, Chap. IV.

If k denotes a different increment of x, then

$$f(x+k)=fx + kf'(x+\phi k),$$

where ϕ is some proper fraction, a function of x and k; so that

$$\frac{f(x+h)-fx}{f(x+k)-fx}=\frac{h}{k}\frac{f'(x+\theta h)}{f'(x+\phi k)}.$$

Ultimately when h and k are sufficiently small

$$\frac{f'(x+\theta h)}{f'(x+\phi k)}=1$$

without sensible error, and then

$$\frac{f(x+h)-fx}{f(x+k)-fx}=\frac{h}{k},$$

which is the rule of *proportional parts*; equivalent to supposing the arc PQ to be straight without introducing sensible error.

11. If x', y' are the co-ordinates of any point P' (figure 3) on the tangent TP,

$$\frac{y'-y}{x'-x}=\frac{dy}{dx},$$

or

$$y'-y=\frac{dy}{dx}(x'-x),$$

the equation of the tangent TP.

TM is called the subtangent, and MG the subnormal at P, PG being the normal at P.

Therefore

$$TM=y\cot\psi=y\frac{dx}{dy},$$

and

$$MG=y\tan\psi=y\frac{dy}{dx}.$$

Also the equation of the normal is

$$y'-y=-\frac{dx}{dy}(x'-x).$$

Instead of the letter h the symbol Δx is often employed to denote the increment of x, and Δy is then used to denote the corresponding increment of y, where y is a function of x, denoted by fx.   Therefore

$$y + \Delta y = f(x + \Delta x),$$

and $$\Delta y = f(x + \Delta x) - fx;$$

so that $$\Delta y = RQ \text{ if } \Delta x = PR;$$

and $$\frac{dy}{dx} = \text{lt}\frac{\Delta y}{\Delta x}.$$

12. Again, let the length of the arc CP of the curve CPQ, measured from any fixed point C to the variable point P, be denoted by s.

Denoting by Δs the increment of s, corresponding to the increment Δx of x, then the arc $CPQ = s + \Delta s$, and the arc $PQ = \Delta s$.

Now, when the point Q approaches to coincidence with P, it is assumed as axiomatic that

$$\text{lt}\frac{\text{chord } PQ}{\text{arc } PQ} = 1\,;$$

or $$\text{lt}\frac{\sqrt{(PR^2 + RQ^2)}}{\text{arc } PQ} = 1\,;$$

or $$\text{lt}\frac{\sqrt{(\Delta x^2 + \Delta y^2)}}{\Delta s} = 1\,;$$

or $$\text{lt}\left(\frac{\Delta x^2}{\Delta s^2} + \frac{\Delta y^2}{\Delta s^2}\right) = 1.$$

Therefore $$\frac{dx^2}{ds^2} + \frac{dy^2}{ds^2} = 1,$$

or, as it is sometimes written,

$$\left(\frac{dx}{ds}\right)^2 + \left(\frac{dy}{ds}\right)^2 = 1.$$

Again, $\cos \psi = \cos xTP$

$$= \mathrm{lt}\, \cos RPQ$$

$$= \mathrm{lt}\, \frac{PR}{PQ}$$

$$= \mathrm{lt}\, \frac{\Delta x}{\Delta s} = \frac{dx}{ds};$$

and $\sin \psi = \sin xTP$

$$= \mathrm{lt}\, \sin RPQ$$

$$= \mathrm{lt}\, \frac{RQ}{PQ}$$

$$= \mathrm{lt}\, \frac{\Delta y}{\Delta s} = \frac{dy}{ds}.$$

Therefore the tangent $PT = y\,\mathrm{cosec}\,\psi = y\dfrac{ds}{dy}$,

and the normal $\qquad PG = y\quad\sec\psi = y\dfrac{ds}{dx}.$

13. The dynamical interpretation of differential co-efficients is interesting; for if x and y, the co-ordinates of a point P moving along the curve CPQ, are given as functions of the time t, then the component velocity of P parallel to the axis Ox

$$= \mathrm{lt}\, \frac{PR}{\Delta t} = \mathrm{lt}\, \frac{\Delta x}{\Delta t} = \frac{dx}{dt};$$

and similarly the component velocity of P parallel to the axis Oy is $\dfrac{dy}{dt}$.

The resultant velocity of the point P is $\dfrac{ds}{dt}$ in the direction of the tangent TP; and therefore

$$\frac{dx^2}{dt^2} + \frac{dy^2}{dt^2} = \frac{ds^2}{dt^2}.$$

14. Since
$$\frac{\Delta y}{\Delta t} = \frac{\Delta y}{\Delta x}\frac{\Delta x}{\Delta t},$$
therefore, proceeding to the limit,
$$\frac{dy}{dt} = \frac{dy}{dx}\frac{dx}{dt},$$
where x is a function of t, and y is a function of x and therefore of t.

With different letters, supposing z is a function of y, and y is a function of x, the *independent* variable, then
$$\frac{dz}{dx} = \frac{dz}{dy}\frac{dy}{dx},$$
the formula required for the *differentiation of a function of a function.*

Thus
$$\frac{dy^m}{dx} = my^{m-1}\frac{dy}{dx},$$

for instance, $\dfrac{d(\sin x)^m}{dx} = m(\sin x)^{m-1}\cos x.$

$$\frac{d\sin y}{dx} = \cos y\frac{dy}{dx},$$

$$\frac{d(\sin y)^m}{dx} = m(\sin y)^{m-1}\cos y\frac{dy}{dx}.$$

$$\frac{d\sin^{-1}y}{dx} = \frac{\dfrac{dy}{dx}}{\sqrt{(1-y^2)}}, \text{ (§ 21).}$$

$$\frac{d\tan^{-1}y}{dx} = \frac{\dfrac{dy}{dx}}{1+y^2}, \text{ (§ 22).}$$

$$\frac{de^y}{dx} = e^y\frac{dy}{dx}, \text{ (§ 26).}$$

$$\frac{d\log y}{dx} = \frac{1}{y}\frac{dy}{dx}, \text{ (§ 26), and so on.}$$

Examples.—Find the d.c. with respect to x of $(\cos x)^n$, $(\tan x)^n$, $(\sec x)^n$, $(\operatorname{cosec} x)^n$, $(\operatorname{vers} x)^n$, $(fx)^n$, $\sin (fx)$.

15. *Implicit and Explicit Relation between two variables, x and y.*

When the variables x and y are connected together by any relation, for instance by

$$x^3 - 3axy + y^3 = 0,$$

this relation is called an *implicit relation* between x and y.

Where however it is possible by solution of the equation to obtain y in terms of x, or x in terms of y, then y is called an *explicit function* of x, or x of y.

For instance, in the above implicit relation between x and y, we cannot obtain y explicitly in terms of x, or x in terms of y, except by the solution of a cubic equation; but from the implicit relation

$$x^2 y^2 - a^2 (x^2 - y^2) = 0,$$

we obtain explicitly

$$y^2 = \frac{a^2 x^2}{a^2 + x^2}, \text{ or } x^2 = \frac{a^2 y^2}{a^2 - y^2}.$$

With an implicit relation between x and y, in order to find $\dfrac{dy}{dx}$, we differentiate the equation with respect to x, treating y as a function of x, and obtain what is called the *first derived equation;* thus from the implicit relation

$$x^3 - 3axy + y^3 = 0,$$

we obtain the first derived equation

$$3x^2 - 3ay - 3ax\frac{dy}{dx} + 3y^2\frac{dy}{dx} = 0,$$

so that

$$\frac{dy}{dx} = \frac{x^2 - ay}{ax - y^2}.$$

B

Examples.—Find $\dfrac{dy}{dx}$ from the implicit relations :—

(1) $ax^2 + 2hxy + by^2 + 2gx + 2fy + c = 0.$

(2) $x^2y^2 = a^3(x + y).$

3) $x^5 + y^5 - 5a^2x^2y = 0.$

(4) Prove that $\dfrac{dy}{dx} = 0$ when $x = a\ \sqrt[3]{2},\ y = a\ \sqrt[3]{4}$;

and $\qquad\qquad \dfrac{dx}{dy} = 0$ when $x = a\ \sqrt[3]{4},\ y = a\ \sqrt[3]{2}$;

if $\qquad\qquad x^3 + y^3 - 3axy = 0.$

16. The student should at this stage be exercised in drawing simple curves from their equations, thus exhibiting to the eye the flow of the function $y = fx.$

The curves should be drawn to scale as carefully as possible; for this purpose logarithm paper ruled into small squares is useful.

Examples.—Draw the curves :

(1) $y = 1,\ x,\ x^2,\ x^3,\ x^4,\ x^n,\ \sqrt{x},\ \dfrac{1}{x},\ \dfrac{1}{x^2}.$

(2) $y^2 = 1,\ x,\ x^2,\ x^3,\ x^4,\ x^n,\ \sqrt{x},\ \dfrac{1}{x},\ \dfrac{1}{x^2}.$

(3) $x^2 - 4x + 2y = 0,\ xy - 2x - y = 0,\ (y - x)^2 = 1 - x^2.$

(4) $y = \dfrac{(x - 1)(x - 2)}{x - 3},\quad y = \dfrac{x - 3}{(x - 1)(x - 2)},\quad y = \dfrac{(x - 1)(x - 3)}{(x - 2)(x - 4)}.$

(5) $x + y = 1,\ x^2 + y^2 = 1,\ x^3 + y^3 = 1,\ \dfrac{1}{x} + \dfrac{1}{y} = 1,\ \dfrac{1}{x^2} + \dfrac{1}{y^2} = 1.$

(6) $x^2 - y^2 = 1,\ y^2 - x^2 = 1,\ \dfrac{1}{x^2} - \dfrac{1}{y^2} = 1,\ \dfrac{1}{y^2} - \dfrac{1}{x^2} = 1.$

(7) $y = \sin x,\ \cos x,\ \tan x,\ \cot x,\ \sec x,\ \text{vers } x,\ x \sin x.$

(8) $y = \sin^{-1}x,\ \cos^{-1}x,\ \tan^{-1}x,\ \cot^{-1}x,\ \sec^{-1}x,\ \text{vers}^{-1}x.$

(9) $y = a^x$, e^x, $\log x$.

(10) Sin $x = \sin y$, $\cos x = \cos y$, $\tan x = \tan y$,
$\sin^2 x + \sin^2 y = 1$, $\tan^2 x + \tan^2 y = 1$.

(11) Prove the properties of the following curves:—

    (i.) In $y = a^x$, the subtangent is constant (§ 26);

    (ii.) In $y^2 = px$, the parabola, the subnormal is constant;

    (iii.) In $x^2 + y^2 = a^2$, the circle, the normal PG is constant;

    (iv.) In $x^{\frac{2}{3}} + y^{\frac{2}{3}} = a^{\frac{2}{3}}$, the part of the tangent intercepted by the axes is constant.

(12) Prove that the equation of the tangent at $(x'y')$

    (i.) Of the circle $x^2 + y^2 = a^2$ is $xx' + yy' = a^2$;

    (ii.) Of the ellipse $\dfrac{x^2}{a^2} + \dfrac{y^2}{b^2} = 1$ is $\dfrac{xx'}{a^2} + \dfrac{yy'}{b^2} = 1$;

    (iii.) Of the parabola $y^2 = px$ is $\dfrac{2y'}{y} - \dfrac{x'}{x} = 1$;

    (iv.) Of the hyperbola $xy = c^2$ is $\dfrac{x'}{x} + \dfrac{y'}{y} = 2$;

    (v.) Of $x^m y^n = c^{m+n}$ is $\dfrac{x'}{nx} + \dfrac{y'}{my} = \dfrac{1}{m} + \dfrac{1}{n}$.

(13) Prove that if

    (i.) $9ay^2 = 4x^3$, $\dfrac{ds}{dx} = \sqrt{\left(1 + \dfrac{x}{a}\right)}$;

    (ii.) $x^{\frac{2}{3}} + y^{\frac{2}{3}} = a^{\frac{2}{3}}$, $\dfrac{ds}{dx} = \left(\dfrac{a}{x}\right)^{\frac{1}{3}}$;

(14) Prove that if

$$\frac{dx}{dt} = hx + by + f, \quad \frac{dy}{dt} = -ax - hy - g,$$

the point P describes the conic

$$ax^2 + 2hxy + by^2 + 2gx + 2fy + c = 0.$$

17. If $OP=r$, and the angle $xOP=\theta$ (fig. 5), then r, θ are called the polar co-ordinates of P.

If $r=f\theta$ is the polar equation of a curve APQ, and if $xOP=\theta$, then $OP=r=f\theta$.

If $\qquad POQ=\Delta\theta$, then $xOQ=\theta+\Delta\theta$;

and $\quad OQ=f(\theta+\Delta\theta)=r+\Delta r$; $\quad RQ=f(\theta+\Delta\theta)-f\theta=\Delta r$;

also $\qquad\qquad\qquad PR=r\Delta\theta,$

if PR is the arc of a circle struck with centre O.

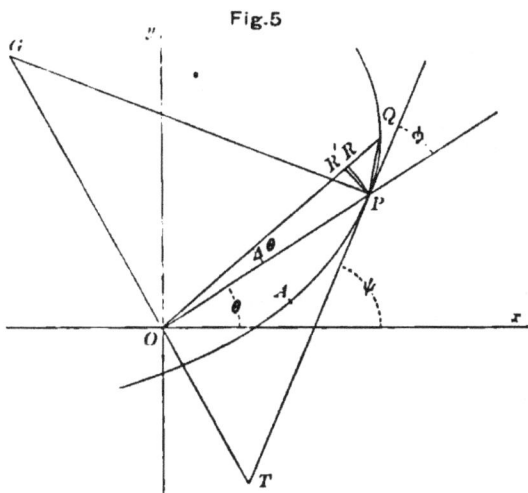

Fig.5

But if PR' is a straight line drawn perpendicular to OQ, then

$$\operatorname{lt}\frac{PR'}{PR}=\operatorname{lt}\frac{\sin\Delta\theta}{\Delta\theta}=1,$$

and $\qquad \operatorname{lt}\frac{RR'}{PR}=\operatorname{lt}\frac{\operatorname{vers}\Delta\theta}{\Delta\theta}=0$; so that $\operatorname{lt}\frac{R'Q}{RQ}=1.$

If the angle between OP and the tangent at P is denoted by ϕ, then

$$\tan \phi = \mathrm{lt} \tan OQP$$
$$= \mathrm{lt} \frac{PR'}{R'Q}$$
$$= \mathrm{lt} \frac{PR}{RQ} \, \mathrm{lt}\frac{PR'}{PR} \, \mathrm{lt}\frac{RQ}{R'Q}$$
$$= \mathrm{lt} \frac{PR}{RQ}$$
$$= \mathrm{lt} \frac{r\Delta\theta}{\Delta r} = \frac{rd\theta}{dr}.$$

18. It the arc AP measured from any fixed point A to a variable point P is denoted by s, and PQ, the increment of the arc, by Δs;

then, since $\qquad \mathrm{lt}\dfrac{\text{chord } PQ}{\text{arc } PQ} = 1,$

therefore $\quad \cos \phi = \mathrm{lt} \cos OQP$
$$= \mathrm{lt}\frac{R'Q}{PQ}$$
$$= \mathrm{lt}\frac{RQ}{\text{arc } PQ} \, \mathrm{lt}\frac{R'Q}{RQ} \, \mathrm{lt}\frac{\text{arc } PQ}{\text{chord } PQ}$$
$$= \mathrm{lt}\frac{RQ}{\text{arc } PQ}$$
$$= \mathrm{lt}\frac{\Delta r}{\Delta s} = \frac{dr}{ds}.$$

Similarly, $\quad \sin \phi = \mathrm{lt} \sin OQP$
$$= \mathrm{lt}\frac{PR'}{P\overline{Q}}$$
$$= \mathrm{lt}\frac{r\Delta\theta}{\Delta s} = \frac{rd\theta}{ds}.$$

Therefore $\quad \dfrac{dr^2}{ds^2} + r^2\dfrac{d\theta^2}{ds^2} = \cos^2\phi + \sin^2\phi = 1.$

If OT is drawn at right angles to OP to meet the tangent at P in T, then OT is called the *polar subtangent*, and

$$OT = r \tan \phi = r^2 \frac{d\theta}{dr}.$$

If TO produced meets the normal at P in G, OG is called the *polar subnormal*, and

$$OG = r \cot \phi = \frac{dr}{d\theta}.$$

The tangent $\qquad PT = r \sec \phi = r \frac{ds}{dr};$

the normal $\qquad PG = r \operatorname{cosec} \phi = \frac{ds}{d\theta}.$

Employing a dynamical interpretation as before, if r and θ, the polar co-ordinates of a point P moving along the curve APQ, are given functions of the time t, then the component velocities of P in the direction OP and in the direction PR', perpendicular to OP, are

$$\frac{dr}{dt} \text{ and } r \frac{d\theta}{dt};$$

these are called the *radial* and *transversal* velocities of P.

The resultant velocity is $\dfrac{ds}{dt}$ in the direction of the tangent TP; and therefore

$$\frac{dr^2}{dt^2} + r^2 \frac{d\theta^2}{dt^2} = \frac{ds^2}{dt^2}.$$

19. *Examples.*—Draw the following curves whose equations are given in polar co-ordinates :—

(1) $r = 1,\ \theta,\ \theta^2,\ \sqrt{\theta},\ \dfrac{1}{\theta}.$

(2) $r = \cos\theta,\ \cos 2\theta,\ \cos 3\theta,\ \cos 4\theta,\ \cos\tfrac{1}{2}\theta,\ \sec\theta,\ \sec\tfrac{1}{2}\theta.$

(3) $r = \sin\theta$, $\sin 2\theta$, $\sin 3\theta$, $\sin 4\theta$, $\sin\frac{1}{2}\theta$, $\operatorname{cosec}\theta$, $\operatorname{cosec}\frac{1}{2}\theta$.

(4) $r = 1 - \cos\theta$, $1 + \cos\theta$, $\dfrac{1}{1 - \cos\theta}$, $\dfrac{1}{1 + \cos\theta}$,

$\dfrac{1}{1 - \cos a\cos\theta}$, $\dfrac{1}{\cos a - \cos\theta}$, $\cos\theta - \cos a$, $1 - \cos a\cos\theta$.

(5) $r = \theta + \sin\theta$, $r = \theta\sin\theta$.

(6) $r = a^\theta$, $\log\theta$.

(7) Prove that—

   (i.) In $r^n = a^n\sin n\theta$, $\phi = n\theta$;

   (ii.) $r^n = b^n\cos n\theta$, $\phi = \frac{1}{2}\pi - n\theta$;

   and prove that these curves cut at right angles.

   (iii.) In $r^n = a^n\sec n\theta$, $\phi = \frac{1}{2}\pi + n\theta$.

   (iv.) $r = a^\theta$, ϕ is constant.

(8) Prove that OT is constant in $r = \dfrac{a}{\theta}$,

            OG „ $r = a\theta$,

            PG „ $r = a\sin\theta$.

(9) Prove that in

   (i.) $r = a(1 + \cos\theta)$, $\dfrac{ds}{d\theta} = 2a\cos\frac{1}{2}\theta$;

   (ii.) $r = a\theta$, $\dfrac{ds}{dr} = \sqrt{\left(1 + \dfrac{r^2}{a^2}\right)}$.

20. It has already been assumed in § 11 that $\dfrac{dy}{dx}$ and $\dfrac{dx}{dy}$ are reciprocal, y being any function of x, and x therefore a function of y.

The proof, if any proof is required, may be given thus: if Δx is any increment of x and Δy the corresponding increment of y, then always

$$\frac{\Delta y}{\Delta x} \times \frac{\Delta x}{\Delta y} = 1;$$

and therefore proceding to the limit,

$$\frac{dy}{dx} \times \frac{dx}{dy} = 1,$$

or $\frac{dy}{dx}$ and $\frac{dx}{dy}$ are reciprocal.

This theorem is required in the

21. *Differentiation of the Inverse Circular Functions.*

Let $\hspace{3cm} y = \sin^{-1}x,$

then $\hspace{2.8cm} x = \sin y,$

$$\frac{dx}{dy} = \cos y$$

$$= \sqrt{(1 - \sin^2 y)}$$

$$= \sqrt{(1 - x^2)} ;$$

$$\frac{dy}{dx} = \frac{1}{\sqrt{(1 - x^2)}},$$

or $\hspace{2cm} \dfrac{d \sin^{-1}x}{dx} = \dfrac{1}{\sqrt{(1 - x^2)}}.$

Similarly $\hspace{1.5cm} \dfrac{d \sin^{-1}\frac{x}{a}}{dx} = \dfrac{1}{\sqrt{(a^2 - x^2)}}.$

. Let $\hspace{2.7cm} y = \cos^{-1}x,$

then $\hspace{2.7cm} x = \cos y,$

$$\frac{dx}{dy} = -\sin y$$

$$= -\sqrt{(1 - x^2)} ;$$

$$\frac{dy}{dx} = -\frac{1}{\sqrt{(1 - x^2)}},$$

or $\hspace{2cm} \dfrac{d \cos^{-1}x}{dx} = -\dfrac{1}{\sqrt{(1 - x^2)}} ;$

and
$$\frac{d\cos^{-1}\frac{x}{a}}{dx} = -\frac{1}{\sqrt{(a^2 - x^2)}}.$$

Since
$$\sin^{-1}\frac{x}{a} + \cos^{-1}\frac{x}{a} = \tfrac{1}{2}\pi,$$

therefore
$$\frac{d\sin^{-1}\frac{x}{a}}{dx} + \frac{d\cos^{-1}\frac{x}{a}}{dx} = 0,$$

which shows why
$$\frac{d\sin^{-1}\frac{x}{a}}{dx} = -\frac{d\cos^{-1}\frac{x}{a}}{dx}.$$

22. Again, if
$$y = \tan^{-1}x,$$
$$x = \tan y,$$
$$\frac{dx}{dy} = \sec^2 y$$
$$= 1 + \tan^2 y$$
$$= 1 + x^2;$$
$$\frac{dy}{dx} = \frac{d\tan^{-1}x}{dx} = \frac{1}{1 + x^2};$$

and
$$\frac{d\tan^{-1}\frac{x}{a}}{dx} = \frac{a}{a^2 + x^2}.$$

If
$$y = \cot^{-1}x,$$
$$x = \cot y,$$
$$\frac{dx}{dy} = -\operatorname{cosec}^2 y$$
$$= -\cot^2 y - 1$$
$$= -x^2 - 1;$$
$$\frac{dy}{dx} = \frac{d\cot^{-1}x}{dx} = -\frac{1}{x^2 + 1};$$

and
$$\frac{d \cot^{-1}\frac{x}{a}}{dx} = -\frac{a}{x^2+a^2}.$$

23. If
$$y = \sec^{-1}x,$$
$$x = \sec y,$$
$$\frac{dx}{dy} = \sec y \tan y$$
$$= x \surd(x^2-1),$$
$$\frac{d \sec^{-1}x}{dx} = \frac{1}{x \surd(x^2-1)}.$$

Similarly
$$\frac{d \cosec^{-1}x}{dx} = -\frac{1}{x \surd(x^2-1)}.$$

If
$$y = \text{vers}^{-1}\frac{x}{a},$$
$$x = a \text{ vers } y = a(1-\cos y),$$
$$\frac{dx}{dy} = a \sin y$$
$$= a \surd\left\{1-\left(1-\frac{x}{a}\right)^2\right\}$$
$$= \surd(2ax-x^2);$$

$$\frac{d \text{ vers}^{-1}\frac{x}{a}}{dx} = \frac{1}{\surd(2ax-x^2)}.$$

24. The inverse circular functions are not required much in Elementary Trigonometry, but are indispensable in the Differential and Integral Calculus, so the principal formulæ of Trigonometry, expressed by the direct functions, are given here, with the corresponding inverse notation.

$\sin(A + B) = \sin A \cos B + \cos A \sin B$;

$\sin^{-1}a + \sin^{-1}b = \sin^{-1}\{a \sqrt{(1 - b^2)} + b \sqrt{(1 - a^2)}\}$.

$\cos(A + B) = \cos A \cos B - \sin A \sin B$;

$\cos^{-1}a + \cos^{-1}b = \cos^{-1}\{ab - \sqrt{(1 - a^2)} \sqrt{(1 - b^2)}\}$.

$\tan(A + B) = \dfrac{\tan A + \tan B}{1 - \tan A \tan B}$;

$\tan^{-1}a + \tan^{-1}b = \tan^{-1}\dfrac{a + b}{1 - ab}$.

$\sin 2A = 2\sin A \cos A$;

$2\sin^{-1}a = \sin^{-1}2a \sqrt{(1 - a^2)}$.

$\cos 2A = 2\cos^2 A - 1$;

$2\cos^{-1}a = \cos^{-1}(2a^2 - 1)$.

$\tan 2A = \dfrac{2\tan A}{1 - \tan^2 A}$, $\sin 2A = \dfrac{2\tan A}{1 + \tan^2 A}$, $\cos 2A = \dfrac{1 - \tan^2 A}{1 + \tan^2 A}$;

$2\tan^{-1}a = \tan^{-1}\dfrac{2a}{1 - a^2} = \sin^{-1}\dfrac{2a}{1 + a^2} = \cos^{-1}\dfrac{1 - a^2}{1 + a^2}$.

$\sin 3A = 3\sin A - 4\sin^3 A$;

$3\sin^{-1}a = \sin^{-1}(3a - 4a^3)$.

$\cos 3A = 4\cos^3 A - 3\cos A$;

$3\cos^{-1}a = \cos^{-1}(4a^3 - 3a)$.

$\tan 3A = \dfrac{3\tan A - \tan^3 A}{1 - 3\tan^2 A}$;

$3\tan^{-1}a = \tan^{-1}\dfrac{3a - a^3}{1 - 3a^2}$.

$\sin 4A = 4\sin A \cos A(1 - 2\sin^2 A)$;

$4\sin^{-1}a = \sin^{-1}4a(1 - 2a^2)\sqrt{(1 - a^2)}$;

and so on.

DIFFERENTIATION.

TABLE I.—GIVING ANY DIRECT CIRCULAR

	Sine.	Cosine.	Tangent.	Cotangent
$\sin\theta =$	$\sin\theta$	$\surd(1-\cos^2\theta)$	$\dfrac{\tan\theta}{\surd(1+\tan^2\theta)}$	$\dfrac{1}{\surd(\cot^2\theta+1)}$
$\cos\theta =$	$\surd(1-\sin^2\theta)$	$\cos\theta$	$\dfrac{1}{\surd(1+\tan^2\theta)}$	$\dfrac{\cot\theta}{\surd(\cot^2\theta+1)}$
$\tan\theta =$	$\dfrac{\sin\theta}{\surd(1-\sin^2\theta)}$	$\dfrac{\surd(1-\cos^2\theta)}{\cos\theta}$	$\tan\theta$	$\dfrac{1}{\cot\theta}$
$\cot\theta =$	$\dfrac{\surd(1-\sin^2\theta)}{\sin\theta}$	$\dfrac{\cos\theta}{\surd(1-\cos^2\theta)}$	$\dfrac{1}{\tan\theta}$	$\cot\theta$
$\sec\theta =$	$\dfrac{1}{\surd(1-\sin^2\theta)}$	$\dfrac{1}{\cos\theta}$	$\surd(1+\tan^2\theta)$	$\dfrac{\surd(\cot^2\theta+1)}{\cot\theta}$
$\operatorname{cosec}\theta =$	$\dfrac{1}{\sin\theta}$	$\dfrac{1}{\surd(1-\cos^2\theta)}$	$\dfrac{\surd(1+\tan^2\theta)}{\tan\theta}$	$\surd(\cot^2\theta+1)$
$\operatorname{vers}\theta =$	$1-\surd(1-\sin^2\theta)$	$1-\cos\theta$	$1-\dfrac{1}{\surd(1+\tan^2\theta)}$	$1-\dfrac{\cot\theta}{\surd(\cot^2\theta+1)}$

FUNCTION IN TERMS OF THE OTHERS.

Secant.	Cosecant.	Versed Sine.	
$\dfrac{\sqrt{(\sec^2\theta-1)}}{\sec\theta}$	$\dfrac{1}{\csc\theta}$	$\sqrt{(2\text{ vers }\theta-\text{vers}^2\theta)}$	$=\sin\theta$
$\dfrac{1}{\sec\theta}$	$\dfrac{\sqrt{(\csc^2\theta-1)}}{\csc\theta}$	$1-\text{vers }\theta$	$=\cos\theta$
$\sqrt{(\sec^2\theta-1)}$	$\dfrac{1}{\sqrt{(\csc^2\theta-1)}}$	$\dfrac{\sqrt{(2\text{ vers }\theta-\text{vers}^2\theta)}}{1-\text{vers }\theta}$	$=\tan\theta$
$\dfrac{1}{\sqrt{(\sec^2\theta-1)}}$	$\sqrt{(\csc^2\theta-1)}$	$\dfrac{1-\text{vers }\theta}{\sqrt{(2\text{ vers }\theta-\text{vers}^2\theta)}}$	$=\cot\theta$
$\sec\theta$	$\dfrac{\csc\theta}{\sqrt{(\csc^2\theta-1)}}$	$\dfrac{1}{1-\text{vers }\theta}$	$=\sec\theta$
$\dfrac{\sec\theta}{\sqrt{(\sec^2\theta-1)}}$	$\csc\theta$	$\dfrac{1}{\sqrt{(2\text{ vers }\theta-\text{vers}^2\theta)}}$	$=\csc\theta$
$1-\dfrac{1}{\sec\theta}$	$1-\dfrac{\sqrt{(\csc^2\theta-1)}}{\csc\theta}$	$\text{vers }\theta$	$=\text{vers }\theta$

TABLE II.—OF THE CORRESPONDING RELATIONS

	Sin^{-1}.	Cos^{-1}.	Tan^{-1}.	Cot^{-1}.
$\sin^{-1}x=$	$\sin^{-1}x$	$\cos^{-1}\sqrt{(1-x^2)}$	$\tan^{-1}\dfrac{x}{\sqrt{(1-x^2)}}$	$\cot^{-1}\dfrac{\sqrt{(1-x^2)}}{x}$
$\cos^{-1}x=$	$\sin^{-1}\sqrt{(1-x^2)}$	$\cos^{-1}x$	$\tan^{-1}\dfrac{\sqrt{(1-x^2)}}{x}$	$\cot^{-1}\dfrac{x}{\sqrt{(1-x^2)}}$
$\tan^{-1}x=$	$\sin^{-1}\dfrac{x}{\sqrt{(1+x^2)}}$	$\cos^{-1}\dfrac{1}{\sqrt{(1+x^2)}}$	$\tan^{-1}x$	$\cot^{-1}\dfrac{1}{x}$
$\cot^{-1}x=$	$\sin^{-1}\dfrac{1}{\sqrt{(x^2+1)}}$	$\cos^{-1}\dfrac{x}{\sqrt{(x^2+1)}}$	$\tan^{-1}\dfrac{1}{x}$	$\cot^{-1}x$
$\sec^{-1}x=$	$\sin^{-1}\dfrac{\sqrt{(x^2-1)}}{x}$	$\cos^{-1}\dfrac{1}{x}$	$\tan^{-1}\sqrt{(x^2-1)}$	$\cot^{-1}\dfrac{1}{\sqrt{(x^2-1)}}$
$\operatorname{cosec}^{-1}x=$	$\sin^{-1}\dfrac{1}{x}$	$\cos^{-1}\dfrac{\sqrt{(x^2-1)}}{x}$	$\tan^{-1}\dfrac{1}{\sqrt{(x^2-1)}}$	$\cot^{-1}\sqrt{(x^2-1)}$
$\operatorname{vers}^{-1}x=$	$\sin^{-1}\sqrt{(2x-x^2)}$	$\cos^{-1}(1-x)$	$\tan^{-1}\dfrac{\sqrt{(2x-x^2)}}{1-x}$	$\cot^{-1}\dfrac{1-x}{\sqrt{(2x-x^2)}}$

OF THE INVERSE CIRCULAR FUNCTIONS.

Sec^{-1}.	Cosec^{-1}.	Vers^{-1}.	
$\sec^{-1}\dfrac{1}{\sqrt{(1-x^2)}}$	$\operatorname{cosec}^{-1}\dfrac{1}{x}$	$\operatorname{vers}^{-1}\left\{1-\sqrt{(1-x^2)}\right\}$	$=\sin^{-1}x$
$\sec^{-1}\dfrac{1}{x}$	$\operatorname{cosec}^{-1}\dfrac{1}{\sqrt{(1-x^2)}}$	$\operatorname{vers}^{-1}(1-x)$	$=\cos^{-1}x$
$\sec^{-1}\sqrt{(1+x^2)}$	$\operatorname{cosec}^{-1}\dfrac{\sqrt{(1+x^2)}}{x}$	$\operatorname{vers}^{-1}\left\{1-\dfrac{1}{\sqrt{(1+x^2)}}\right\}$	$=\tan^{-1}x$
$\sec^{-1}\dfrac{\sqrt{(x^2+1)}}{x}$	$\operatorname{cosec}^{-1}\sqrt{(x^2+1)}$	$\operatorname{vers}^{-1}\left\{1-\dfrac{x}{\sqrt{(x^2+1)}}\right\}$	$=\cot^{-1}x$
$\sec^{-1}x$	$\operatorname{cosec}^{-1}\dfrac{x}{\sqrt{(x^2-1)}}$	$\operatorname{vers}^{-1}\left(1-\dfrac{1}{x}\right)$	$=\sec^{-1}x$
$\sec^{-1}\dfrac{x}{\sqrt{(x^2-1)}}$	$\operatorname{cosec}^{-1}x$	$\operatorname{vers}^{-1}\left\{1-\dfrac{\sqrt{(x^2-1)}}{x}\right\}$	$=\operatorname{cosec}^{+1}x$
$\sec^{-1}\dfrac{1}{1-x}$	$\operatorname{cosec}^{-1}\dfrac{1}{\sqrt{(2x-x^2)}}$	$\operatorname{vers}^{-1}x$	$=\operatorname{vers}^{-1}x$

Since all the circular functions are expressible in terms of any one of them, we might use only one function, say the *tangent*, and its inverse; but the other functions are used in order to avoid irrational and complicated expressions.

25. The d.c.'s of some of the inverse functions can be obtained almost as simply by the direct process; thus:

$$\frac{d \tan^{-1}x}{dx} = \mathrm{lt}\ \frac{\tan^{-1}(x+h) - \tan^{-1}x}{h}$$

$$= \mathrm{lt}\ \frac{1}{h}\tan^{-1}\frac{x+h-x}{1+x(x+h)}$$

$$= \mathrm{lt}\ \frac{1}{h}\tan^{-1}\frac{h}{1+x^2+xh}$$

$$= \frac{1}{1+x^2};$$

because $\dfrac{\tan^{-1}z}{z} = 1$ when $z = 0$ (§ 5), and

here $z = \dfrac{h}{1+x^2+xh}.$

By the direct process

$$\frac{d \sin^{-1}x}{dx} = \mathrm{lt}\ \frac{\sin^{-1}(x+h) - \sin^{-1}x}{h}$$

$$= \mathrm{lt}\ \frac{\sin^{-1}\{(x+h)\sqrt{(1-x^2)} - x\sqrt{(1-x^2-2xh-h^2)}\}}{h}.$$

Denoting $(x+h)\sqrt{(1-x^2)} - x\sqrt{(1-x^2-2xh-h^2)}$ by z, then $z = 0$ when $h = 0$; and therefore

$$\frac{d \sin^{-1}x}{dx} = \mathrm{lt}\,\frac{\sin^{-1}z}{h} = \mathrm{lt}\,\frac{z}{h} \quad (\S\ 5).$$

$$= \mathrm{lt}\,\frac{(x+h)\sqrt{(1-x^2)} - x\sqrt{(1-x^2-2xh-h^2)}}{h}$$

$$= \mathrm{lt}\,\frac{(x+h)^2(1-x^2) - x^2(1-x^2-2xh-h^2)}{h\{(x+h)\sqrt{(1-x^2)} + x\sqrt{(1-x^2-2xh-h^2)}\}}$$

$$= \mathrm{lt}\,\frac{2x+h}{(x+h)\sqrt{(1-x^2)} + x\sqrt{(1-x^2-2xh-h^2)}}$$

$$= \frac{1}{\sqrt{(1-x^2)}}.$$

Example.—Determine in the same way the d.c. with respect to x of $\sin^{-1}\frac{x}{a}$, $\cos^{-1}\frac{x}{a}$, $\tan^{-1}\frac{x}{a}$, $\sec^{-1}\frac{x}{a}$, $\mathrm{vers}^{-1}\frac{x}{a}$.

26. Before $\dfrac{da^x}{dx}$ and $\dfrac{d \log_a x}{dx}$ can be found, the number e, the *base* of the *natural* or *Napierian* logarithms, must be defined.

When $m = \infty$, $\left(1 + \dfrac{1}{m}\right)^m = (1+0)^\infty$, which is *indeterminate*, and its value must therefore be found by the *method of limits.*

Expanding by the binomial theorem,

$$\mathrm{lt}\,\left(1 + \frac{1}{m}\right)^m$$

$$= \mathrm{lt}\,\left\{ 1 + m\frac{1}{m} + \frac{m(m-1)}{2!}\frac{1}{m^2} + \frac{m(m-1)(m-2)}{3!}\frac{1}{m^3} + \cdots \right\}$$

$$= \mathrm{lt}\,\left\{ 1 + \frac{1}{1!} + \frac{1-\dfrac{1}{m}}{2!} + \frac{\left(1-\dfrac{1}{m}\right)\left(1-\dfrac{2}{m}\right)}{3!} + \cdots \right\}$$

$$= 1 + \frac{1}{1!} + \frac{1}{2!} + \frac{1}{3!} + \cdots \text{ to infinity}$$

$$
\begin{aligned}
= \quad & 1 \\
+ \; & 1 \\
+ \; & \cdot 5 \\
+ \; & \cdot 166666\ldots \\
+ \; & \cdot 041666\ldots \\
+ \; & \cdot 008333\ldots \\
+ \; & \cdot 001388\ldots \\
+ \; & \cdot 000198\ldots \\
+ \; & \cdot 000024\ldots \\
+ \; & \cdot 000002\ldots \\
+ \; & \ldots\ldots\ldots\ldots \\
= \quad & 2\cdot71828\ldots.
\end{aligned}
$$

$$\ldots\ldots\ldots(1),$$

an incommensurable number, denoted by the letter e, and called the *base* of the natural or Napierian logarithms.

Next put $m = \dfrac{1}{z}$; then when $m = \infty$, $z = 0$; and therefore

$$e = \mathrm{lt}\left(1 + \frac{1}{m}\right)^{m} \text{ when } m = \infty$$

$$= \mathrm{lt}(1 + z)^{\frac{1}{z}} \text{ when } z = 0 \ldots\ldots\ldots\ldots\ldots(2).$$

Taking logarithms to any base a,

$$\log_a e = \mathrm{lt}\, \log_a(1 + z)^{\frac{1}{z}}$$

$$= \mathrm{lt}\, \frac{\log_a(1 + z)}{z} \text{ when } z = 0 \ldots\ldots(3),$$

and $\qquad \mathrm{lt}\, \dfrac{\log_e(1 + z)}{z} = \log_e e = 1.$

Next put $\qquad \log_a(1 + z) = h,$

then $\qquad\qquad 1 + z = a^{h},$

$$z = a^{h} - 1,$$

and when $z = 0$, $h = 0$.

Therefore $\qquad \log_a e = \mathrm{lt}\, \dfrac{h}{a^{h} - 1};$

and therefore $\text{lt } \dfrac{a^h - 1}{h} = \dfrac{1}{\log_a e}$

$$= \log_e a \dots\dots\dots\dots\dots\dots\dots\dots\dots(4).$$

and $\text{lt } \dfrac{e^h - 1}{h} = \log_e e = 1.$

Therefore $\dfrac{da^x}{dx} = \text{lt } \dfrac{a^{x+h} - a^x}{h}$

$$= a^x \text{lt } \dfrac{a^h - 1}{h} = a^x \log_e a,$$

and $\dfrac{de^x}{dx} = e^x \log_e e = e^x.$

Also $\dfrac{d \log_a x}{dx} = \text{lt } \dfrac{\log_a (x + h) - \log_a x}{h}$

$$= \text{lt } \dfrac{\log_a \dfrac{x + h}{x}}{h}$$

$$= \dfrac{1}{x} \text{lt } \dfrac{\log_a \left(1 + \dfrac{h}{x}\right)}{\dfrac{h}{x}}$$

$$= \dfrac{1}{x} \text{lt } \dfrac{\log_a (1 + z)}{z}, \text{ if } \dfrac{h}{x} = z,$$

$$= \dfrac{1}{x} \log_a e.$$

Therefore $\dfrac{d \log x}{dx} = \dfrac{1}{x},$

when the base is e. The logarithms to the base e are called *natural* logarithms: natural logarithms are intended when the base is not indicated.

The circular measure will also be invariably adopted for the measurement of angles.

By these means extraneous factors will be found to disappear in differentiation, and the expressions will be much simplified.

For instance, $\dfrac{d \sin x°}{dx} = \dfrac{\pi}{180} \cos x°$.

27. The *exponential* function a^x and the *logarithmic* function $\log_a x$ are inverse functions, because if

$$a^m = p, \text{ then } m = \log_a p.$$

Starting with $m = \log_a p$, then p is sometimes denoted by $\exp_a m$ or $\log^{-1}_a m$, instead of a^m, especially when m is a complicated expression.

By the theory of indices as explained in Algebra, if $a^n = q$, then

$$pq = a^m a^n = a^{m+n} \dots\dots\dots\dots\dots\dots(1)$$

$$\frac{p}{q} = \frac{a^m}{a^n} = a^{m-n} \dots\dots\dots\dots\dots\dots(2)$$

$$p^r = (a^m)^r = a^{mr} \dots\dots\dots\dots\dots\dots(3)$$

$$\sqrt[r]{p} = (a^m)^{\frac{1}{r}} = a^{\frac{m}{r}} \dots\dots\dots\dots\dots\dots(4)$$

Therefore the corresponding theorems for the logarithmic function are:

$$\log_a pq = m + n = \log_a p + \log_a q \dots\dots\dots\dots(5)$$

$$\log \frac{p}{q} = m - n = \log p - \log q \dots\dots\dots\dots(6)$$

$$\log p^r = mr = r \log p \dots\dots\dots\dots(7)$$

$$\log \sqrt[r]{p} = \frac{m}{r} = \frac{1}{r} \log p \dots\dots\dots\dots(8)$$

Also if $e^b = a$, then $e = a^{\frac{1}{b}}$, and therefore $b = \log_e a$, and $\dfrac{1}{b} = \log_a e$; so that $\log_e a \log_a e = 1$.

These theorems have already been employed in establishing the preceding differentiations.

28. Corresponding to the trigonometrical functions of the circle, there are certain functions associated with the hyperbola, called the *hyperbolic functions*, which are of very great use, and are defined here as follows:

$\frac{1}{2}(e^v - e^{-v})$ is called the *hyperbolic sine* of v, and is denoted by $\sinh v$;

$\frac{1}{2}(e^v + e^{-v})$ is called the *hyperbolic cosine* of v, and denoted by $\cosh v$; and

$\dfrac{e^v - e^{-v}}{e^v + e^{-v}} = \dfrac{\sinh v}{\cosh v}$ is called the *hyperbolic tangent* of v,

and is denoted by $\tanh v$; and so on for the other hyperbolic functions corresponding to the remaining circular functions.

From the exponential values of the sine and cosine, i denoting $\sqrt{-1}$, namely

$$\sin v = \frac{1}{2i}(e^{iv} - e^{-iv}),\ \cos v = \frac{1}{2}(e^{iv} + e^{-iv});$$

it is seen that

$$\sin iv = i \sinh v,\ \cos iv = \cosh v;$$
$$\cosh^2 v - \sinh^2 v = 1;$$

and
$$\sin(u + iv) = \sin u \cosh v + i \cos u \sinh v,$$
$$\cos(u + iv) = \cos u \cosh v - i \sin u \sinh v.$$

29. For instance, if
$$x + iy = c \cos(\xi - i\eta),$$
then
$$x = c \cos \xi \cosh \eta,$$
$$y = c \sin \xi \sinh \eta,$$
and alternately eliminating ξ and η,
$$\frac{x^2}{c^2 \cosh^2 \eta} + \frac{y^2}{c^2 \sinh^2 \eta} = 1,$$
$$\frac{x^2}{c^2 \cos^2 \xi} - \frac{y^2}{c^2 \sin^2 \xi} = 1;$$

representing a system of confocal ellipses and hyperbolas (fig. 6) for constant values of η and ξ; and sech η will be the excentricity of an ellipse, sec ξ of a hyperbola.

Fig.6

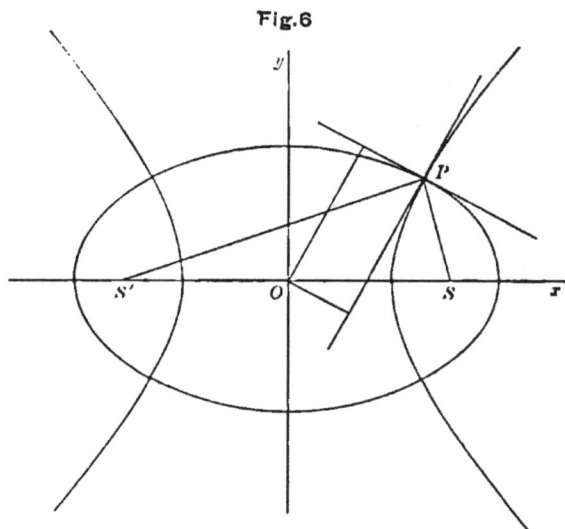

Again, if $\quad x+iy=c\,\tan\tfrac{1}{2}(\chi+i\rho)\,;$

this may be written

$$x+iy=c\frac{\sin\tfrac{1}{2}(\chi+i\rho)\cos\tfrac{1}{2}(\chi-i\rho)}{\cos\tfrac{1}{2}(\chi+i\rho)\cos\tfrac{1}{2}(\chi-i\rho)}=c\frac{\sin\chi+\sin i\rho}{\cos i\rho+\cos\chi}$$

so that $\quad\quad x=\dfrac{c\sin\chi}{\cosh\rho+\cos\chi},$

$$y=\dfrac{c\sinh\rho}{\cosh\rho+\cos\chi}.$$

Also $\quad\tan\chi=\tan\{\tfrac{1}{2}(\chi+i\rho)+\tfrac{1}{2}(\chi-i\rho)\}$

$$=c\frac{x+iy+x-iy}{c^2-(x+iy)(x-iy)}$$

$$=\frac{2cx}{c^2-x^2-y^2},$$

or $\quad\quad\quad x^2+y^2+2cx\cot\chi-c^2=0\,;$

representing a circle of radius $c \operatorname{cosec} \chi$ and centre $(-c \cot \chi, 0)$, and passing through $(0, \pm c)$.

And
$$\tan i\rho = \tan\{\tfrac{1}{2}(\chi + i\rho) - \tfrac{1}{2}(\chi - i\rho)\}$$

$$= c\frac{x + iy - x + iy}{c^2 + (x + iy)(x - iy)}$$

$$= \frac{2icy}{c^2 + x^2 + y^2},$$

or
$$\tanh\rho = \frac{2cy}{c^2 + x^2 + y^2},$$

or
$$x^2 + y^2 - 2cy \coth \rho + c^2 = 0 ;$$

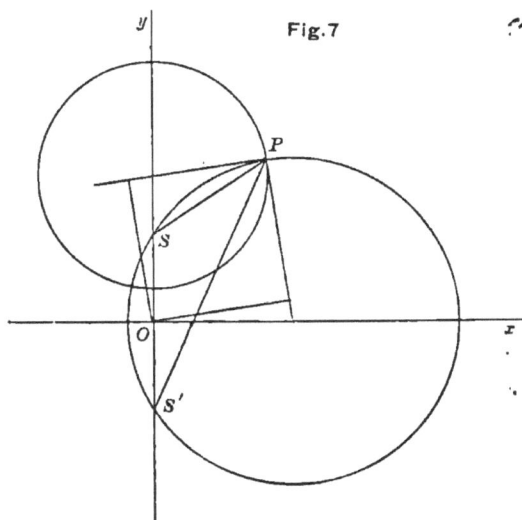

Fig.7

representing a circle of radius $c \operatorname{cosech} \rho$ and centre $(0, c \coth \rho)$, which always cuts the preceding circle (fig. 7) at right angles.

As an exercise, prove that the focal distances of a point

at the intersection of the ellipse η and the hyperbola ξ are

$$c(\cosh \eta - \cos \xi) \text{ and } c(\cosh \eta + \cos \xi),$$

and that the perpendiculars from the centre on the tangents are

$$\frac{c \sinh 2\eta}{\sqrt{\{2(\cosh 2\eta - \cos 2\xi)\}}} \text{ and } -\frac{c \sin 2\xi}{\sqrt{\{2(\cosh 2\eta - \cos 2\xi)\}}}.$$

Determine also for the orthogonal circles of fig. 7 the distances of any point from S and S', and the perpendiculars on the tangents in terms of χ and ρ; and prove that χ is the angle between SP and $S'P$, and $\rho = \log \dfrac{S'P}{SP}$.

30. The geometrical interpretation of the hyperbolic functions can be seen by drawing (Fig. 8) the rectangular hyperbola

$$x^2 - y^2 = a^2, \dots\dots\dots\dots\dots\dots(1)$$

the conjugate hyperbola

$$y^2 - x^2 = a^2, \dots\dots\dots\dots\dots\dots(2)$$

and the inscribed circle

$$x^2 + y^2 = a^2 \dots\dots\dots\dots \dots\dots\dots(3)$$

Then if P is any point (xy) on (1), we can put

$$x = OM = a \cosh u,$$
$$y = MP = a \sinh u,$$

since $\cosh^2 u - \sinh^2 u = 1$;

and $\dfrac{y}{x} = \tan xOP = \tanh u.$

Then the area AOP will be found to be $\frac{1}{2}a^2 u$ (Chap. II.).

From M draw the tangent MR to the circle (3); then if the circular measure of xOR is θ, the sectorial area AOR is $\frac{1}{2}a^2\theta$; and

$$x = a \sec \theta = a \cosh u,$$
$$y = a \tan \theta = a \sinh u,$$

so that $MR = MP.$

When θ and u are connected by this relation, θ is called by Professor Cayley the *Gudermannian* of u (from Gudermann who studied the properties of these functions),

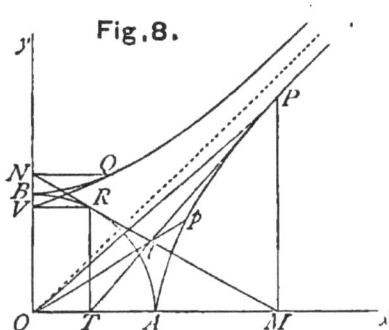

Fig.8.

and is denoted by gd u; and $\sin \theta$ by sg u, $\cos \theta$ by cg u, etc.; so that sg $u = \tanh u$, cg $u = \operatorname{sech} u$, etc.

It may be noticed that, if

$$\cos \theta \cosh u = 1,$$

then

$$\frac{1 - \tan^2 \tfrac{1}{2}\theta}{1 + \tan^2 \tfrac{1}{2}\theta} \times \frac{1 + \tanh^2 \tfrac{1}{2}u}{1 - \tanh^2 \tfrac{1}{2}u} = 1,$$

or

$$\tan^2 \tfrac{1}{2}\theta = \tanh^2 \tfrac{1}{2}u,$$

so that the line Otp which bisects the angle AOR bisects also the sectorial area AOP; and Otp, MR, and the tangent at P will be found to intersect in one point on the tangent at A.

(When the theory of *Elliptic Functions* is studied it will be found that the circular functions are a particular case of the elliptic functions obtained by making the *modulus* of the elliptic functions zero, and the hyperbolic or Gudermannian functions a particular case obtained by making the modulus unity.)

31. If x, y are the co-ordinates of any point Q on the conjugate hyperbola (2), then we may put

$$x = NQ = a \sinh v,$$
$$y = ON = a \cosh v;$$

and then the sectorial area OBQ will be found to be $\frac{1}{2}a^2 v$; and $u = v$, when OP and OQ are conjugate.

Generally, by orthogonal projection on a plane through the transverse axis, for any point P on the hyperbola

$$\frac{x^2}{a^2} - \frac{y^2}{b^2} = 1,$$

we have $\qquad x = a \cosh u, \ y = b \sinh u,$

where the sectorial area AOP

$$= \tfrac{1}{2}abu = \tfrac{1}{2}ab \log\left(\frac{x}{a} + \frac{y}{b}\right);$$

and for any point Q on the conjugate hyperbola

$$\frac{x^2}{a^2} - \frac{y^2}{b^2} = -1,$$

$$x = a \sinh v, \ y = b \cosh v,$$

where the sectorial area BOQ

$$= \tfrac{1}{2}abv = \tfrac{1}{2}ab \log\left(\frac{x}{a} + \frac{y}{b}\right);$$

and $u = v$, when OP and OQ are conjugate.

32. Corresponding to the trigonometrical formulæ for the circular functions, we have for the hyperbolic functions

$$\cosh^2 u - \sinh^2 u = 1;$$
$$\tanh^2 u + \operatorname{sech}^2 u = 1;$$
$$\coth^2 u - \operatorname{cosech}^2 u = 1;$$
$$\sinh 2u = 2\sinh u \cosh u;$$

$$\cosh 2u = \cosh^2 u + \sinh^2 u,$$
$$= 2\cosh^2 u - 1,$$
$$= 1 + 2\sinh^2 u;$$
$$\sinh(u+v) = \sinh u \cosh v + \cosh u \sinh v;$$
$$\cosh(u+v) = \cosh u \cosh v + \sinh u \sinh v;$$
$$\tanh(u+v) = \frac{\tanh u + \tanh v}{1 + \tanh u \tanh v};$$
$$\sinh u + \sinh v = 2 \sinh \tfrac{1}{2}(u+v)\cosh \tfrac{1}{2}(u-v);$$
$$\sinh u - \sinh v = 2 \cosh \tfrac{1}{2}(u+v)\sinh \tfrac{1}{2}(u-v);$$
$$\cosh u + \cosh v = 2 \cosh \tfrac{1}{2}(u+v)\cosh \tfrac{1}{2}(u-v);$$
$$\cosh u - \cosh v = 2 \sinh \tfrac{1}{2}(u+v)\sinh \tfrac{1}{2}(u-v);$$
$$\tanh u - \tanh v = \sinh(u-v)\operatorname{sech} u \operatorname{sech} v.$$

Also
$$\frac{d \sinh x}{dx} = \cosh x,$$

$$\frac{d \cosh x}{dx} = \sinh x,$$

$$\frac{d \tanh x}{dx} = \operatorname{sech}^2 x,$$

$$\frac{d \coth x}{dx} = -\operatorname{cosech}^2 x;$$

which may be proved in a manner corresponding to that employed for the circular functions; because

$$\frac{\sinh h}{h} = 1, \text{ and } \frac{\tanh h}{h} = 1, \text{ when } h = 0.$$

Corresponding to the *real* period 2π or π of the circular functions there is an *imaginary* period $2i\pi$ of the hyperbolic sine and cosine, and an imaginary period $i\pi$ of the hyperbolic tangent.

Thus,
$$\sinh \tfrac{1}{2}i\pi = i, \ \cosh \tfrac{1}{2}i\pi = 0, \ \tanh \tfrac{1}{2}i\pi = \infty;$$
$$\sinh i\pi = 0, \ \cosh i\pi = -1, \ \tanh i\pi = 0;$$
$$\sinh 2i\pi = 0, \ \cosh 2i\pi = 1, \ \tanh 2i\pi = 0;$$

$$\sinh(\tfrac{1}{2}i\pi + v) = i\cosh v,\ \cosh(\tfrac{1}{2}i\pi + v) = i\sinh v,$$
$$\tanh(\tfrac{1}{2}i\pi + v) = \coth v\ ;$$
$$\sinh(i\pi + v) = -\sinh v,\ \cosh(i\pi + v) = -\cosh v,$$
$$\tanh(i\pi + v) = \tanh v\ ;$$
$$\sinh(2i\pi + v) = \sinh v,\ \cosh(2i\pi + v) = \cosh v,$$
$$\tanh(2i\pi + v) = \tanh v.$$

33. The *inverse hyperbolic* functions will be required in the Integral Calculus; the preceding formulæ with the inverse notation become

$$\sinh^{-1}x = \cosh^{-1}\sqrt{(1+x^2)} = \tanh^{-1}\frac{x}{\sqrt{(1+x^2)}} = \dots\ ;$$
$$2\sinh^{-1}x = \sinh^{-1}2x\sqrt{(1+x^2)} = \cosh^{-1}(1+2x^2)\ ;$$
$$2\cosh^{-1}x = \sinh^{-1}2x\sqrt{(x^2-1)} = \cosh^{-1}(2x^2-1)\ ;$$
$$\sinh^{-1}x + \sinh^{-1}y = \sinh^{-1}\{x\sqrt{(1+y^2)} + y\sqrt{(1+x^2)}\};$$
$$\cosh^{-1}x + \cosh^{-1}y = \cosh^{-1}\{xy + \sqrt{(x^2-1)}\sqrt{(y^2-1)}\};$$
$$\tanh^{-1}x + \tanh^{-1}y = \tanh^{-1}\frac{x+y}{1+xy}.$$

It is an instructive exercise to construct, like Tables I. and II. of the circular functions, the corresponding tables of the direct and inverse hyperbolic functions.

The inverse hyperbolic functions can all be expressed by logarithms; for instance, if

$$y = \sinh^{-1}x,\ \text{then}\ x = \sinh y,$$

and

$$\sqrt{(1+x^2)} + x = \cosh y + \sinh y = \exp y,$$

and therefore,

$$y = \sinh^{-1}x = \log\{\sqrt{(1+x^2)} + x\}.$$

Similarly $\cosh^{-1}x = \log\{x + \sqrt{(x^2-1)}\}\ ;$

$$\tanh^{-1}x = \tfrac{1}{2}\log\frac{1+x}{1-x},\ \coth^{-1}x = \tfrac{1}{2}\log\frac{x+1}{x-1}.$$

The use of the inverse hyperbolic functions will be apparent when we investigate their d.c.'s, and compare them with those of the inverse circular functions.

For let
$$y = \sinh^{-1}\frac{x}{a},$$

then
$$x = a \sinh y,$$

$$\frac{dx}{dy} = a \cosh y$$

$$= a \sqrt{(1 + \sinh^2 y)}$$

$$= a \sqrt{\left(1 + \frac{x^2}{a^2}\right)}$$

$$= \sqrt{(a^2 + x^2)} ;$$

$$\frac{dy}{dx} = \frac{d \sinh^{-1}\frac{x}{a}}{dx} = \frac{1}{\sqrt{(a^2 + x^2)}}.$$

Similarly it can be proved that

$$\frac{d \cosh^{-1}\frac{x}{a}}{dx} = \frac{1}{\sqrt{(x^2 - a^2)}} ;$$

$$\frac{d \tanh^{-1}\frac{x}{a}}{dx} = \frac{a}{a^2 - x^2}, \quad x < a ;$$

$$\frac{d \coth^{-1}\frac{x}{a}}{dx} = -\frac{a}{x^2 - a^2}, \quad x > a.$$

34. *Differentiation of the* (i.) *sum* $u + v$ *or difference* $u - v$, (ii.) *product* uv, (iii.) *quotient* $\frac{u}{v}$, *of* u *and* v, *two given functions of* x.

Denoting the increment of x by Δx, and the corresponding increments of u and v by Δu and Δv, then

(i.)
$$\frac{d(u + v)}{dx} = \mathrm{lt}\frac{\Delta u + \Delta v}{\Delta x}$$

$$= \mathrm{lt}\left(\frac{\Delta u}{\Delta x} + \frac{\Delta v}{\Delta x}\right)$$

$$= \frac{du}{dx} + \frac{dv}{dx}.$$

Similarly $$\frac{d(u-v)}{dx} = \frac{du}{dx} - \frac{dv}{dx}.$$

And generally, if a and b denote constant factors,

$$\frac{d(au+bv)}{dx} = a\frac{du}{dx} + b\frac{dv}{dx}.$$

Thus

$$\frac{d}{dx}(ax^m + bx^n - cx^v \ldots\ldots)$$

$$= amx^{m-1} + bnx^{n-1} - cpx^{p-1} \ldots\ldots$$

(ii.) $$\frac{duv}{dx} = \text{lt}\frac{(u+\Delta u)(v+\Delta v) - uv}{\Delta x}$$

$$= \text{lt}\left\{ \frac{\Delta u}{\Delta x}v + (u+\Delta u)\frac{\Delta v}{\Delta x} \right\}$$

$$= \frac{du}{dx}v + u\frac{dv}{dx};$$

because $\text{lt}\dfrac{\Delta u}{\Delta x} = \dfrac{du}{dx}$, $\text{lt}\dfrac{\Delta v}{\Delta x} = \dfrac{dv}{dx}$, and $\text{lt}(u+\Delta u) = u$.

Hence the rule for differentiating a product: *Differentiate with respect to each factor and add the results.*

Thus $$\frac{dx^n\log x}{dx} = nx^{n-1}\log x + x^n\frac{1}{x}$$

$$= x^{n-1}(n\log x + 1)$$

$$= x^{n-1}\log ex^n.$$

(iii) $$\frac{d\frac{u}{v}}{dx} = \text{lt}\frac{\dfrac{u+\Delta u}{v+\Delta v} - \dfrac{u}{v}}{\Delta x}$$

$$= \text{lt}\frac{(u+\Delta u)v - u(v+\Delta v)}{\Delta x(v+\Delta v)v}$$

$$= \text{lt}\frac{\dfrac{\Delta u}{\Delta x}v - u\dfrac{\Delta v}{\Delta x}}{(v+\Delta v)v}$$

$$= \frac{\dfrac{du}{dx}v - u\dfrac{dv}{dx}}{v^2};$$

hence the d.c. of a fraction is another fraction of which *the denominator is the square of the original denominator, and the numerator is the d.c. of the numerator multiplied by the denominator, minus the numerator multiplied by the d.c. of the denominator.*

Thus if
$$y = \frac{3x - x^3}{1 - 3x^2}$$

$$\frac{dy}{dx} = \frac{(3 - 3x^2)(1 - 3x^2) - (3x - x^3)(-6x)}{(1 - 3x^2)^2}$$

$$= 3\left(\frac{1 + x^2}{1 - 3x^2}\right)^2.$$

If the denominator of the fraction is of the form v^n, for instance, if

$$y = \frac{u}{v^n},$$

then
$$\frac{dy}{dx} = \frac{\frac{du}{dx}v^n - unv^{n-1}\frac{dv}{dx}}{v^{2n}}$$

$$= \frac{\frac{du}{dx}v - nu\frac{dv}{dx}}{v^{n+1}},$$

removing the common factor v^{n-1} from the numerator and denominator.

In such cases it is sometimes preferable to write the fraction in the form of a product,

$$y = uv^{-n},$$

then
$$\frac{dy}{dx} = \frac{du}{dx}v^{-n} - nuv^{-n-1}\frac{dv}{dx}$$

$$= \frac{\frac{du}{dx}v - nu\frac{dv}{dx}}{v^{n+1}}$$

as before.

Examples.—Differentiate with respect to x:

$$(1+2x-4x^2)(1-2x+4x^2-4x^3),\ x^2(1+x)^3(2-x)^4,$$

$$\frac{1-x}{x},\ \frac{1+3x^2-3x^2}{3(1-x)^3},\ (x-a)^m(x-b)^n,\ \frac{1}{(x-a)^m(x-b)^n}$$

$$\sqrt{\left(\frac{1-x}{1+x}\right)},\ \cos nx(\sin x)^n,\ n\sin mx\cosh nx-m\cos mx\sinh nx.$$

35. When the function of x to be differentiated is a single term consisting of factors raised to different powers, it is often simpler to take logarithms before differentiating : thus, if

$$y^2=\frac{x^m}{(1+x)^n}$$

then

$$2\log y=m\log x-n\log(1+x),$$

and

$$\frac{2}{y}\frac{dy}{dx}=\frac{m}{x}-\frac{n}{1+x}=\frac{m+(m-n)x}{x(1+x)}\ ;$$

$$\frac{dy}{dx}=\tfrac{1}{2}\frac{m+(m-n)x}{(1+x)^{\frac{1}{2}n+1}}x^{\frac{1}{2}m-1}\ ;$$

this is called *logarithmic differentiation.*

Again if $\quad y=u^v$,

where u and v are functions of x, then

$$\log y=v\log u,$$

and differentiating with respect to x

$$\frac{1}{y}\frac{dy}{dx}=\frac{dv}{dx}\log u+\frac{v}{u}\frac{du}{dx},$$

or

$$\frac{dy}{dx}=u^v\left(\frac{dv}{dx}\log u+\frac{v}{u}\frac{du}{dx}\right).$$

Examples.—Differentiate logarithmically

$$\sqrt{\left(\frac{1-x}{1+x}\right)},\ \frac{(x+1)^{\frac{1}{2}}(x+3)^{\frac{2}{3}}}{(x+2)^4},\ \frac{1}{(x-a)^m(x-b)^n},\ \frac{(\sin mx)^n}{(\cos nx)^m}.$$

$$uvw\ldots,\ x^x,\ e^{x^x}.$$

General Examples of Differentiation.

In the following examples y is given as a function of x, and it is required to determine $\frac{dy}{dx}$ according to the rules explained in this Chapter.

1. $y = \sqrt{(2x)}, \frac{dy}{dx} = \frac{1}{\sqrt{(2x)}}.$

Here $y = (2x)^{\frac{1}{2}}$, and therefore

$$\frac{dy}{dx} = \tfrac{1}{2} \times (2x)^{-\frac{1}{2}} \times 2 = (2x)^{-\frac{1}{2}} = \frac{1}{\sqrt{(2x)}}.$$

2. $y = (nx)^{\frac{1}{n}}, \frac{dy}{dx} = (nx)^{\frac{1}{n}-1}.$

3. $y = \sqrt{(x+a)}, \frac{dy}{dx} = \frac{1}{2\sqrt{(x+a)}}.$

4. $y = \frac{1}{\sqrt{(a-x)}}, \frac{dy}{dx} = \frac{1}{2(a-x)^{\frac{3}{2}}}.$

5. $y = \sqrt{(a^2+x^2)}, \frac{dy}{dx} = \frac{x}{\sqrt{(a^2+x^2)}}.$

6. $y = \frac{1}{\sqrt{(a^2-x^2)}}, \frac{dy}{dx} = \frac{x}{(a^2-x^2)^{\frac{3}{2}}}.$

7. $y = \frac{x}{\sqrt{(x^2+a^2)}}, \frac{dy}{dx} = \frac{a^2}{(x^2+a^2)^{\frac{3}{2}}}.$

Employing the rule of § 34 for the differentiation of a fraction,

$$\frac{dy}{dx} = \frac{\sqrt{(x^2+a^2)} - x\dfrac{x}{\sqrt{(x^2+a^2)}}}{x^2+a^2} = \frac{a^2}{(x^2+a^2)^{\frac{3}{2}}}.$$

8. $y = \sqrt{(2ax-x^2)}, \frac{dy}{dx} = \frac{a-x}{\sqrt{(2ax-x^2)}}.$

D

9. $y = \frac{1}{2}x - \frac{1}{4}\sin 2x$, $\frac{dy}{dx} = \sin^2 x$.

10. $y = \frac{1}{2}x + \frac{1}{4}\sin 2x$, $\frac{dy}{dx} = \cos^2 x$.

11. $y = \frac{1}{3}\cos^3 x - \cos x$, $\frac{dy}{dx} = \sin^3 x$.

12. $y = \sin x - \frac{1}{3}\sin^3 x$, $\frac{dy}{dx} = \cos^3 x$.

13. $y = \frac{3}{8}x + \frac{1}{4}\sin 2x + \frac{1}{32}\sin 4x$, $\frac{dy}{dx} = \cos^4 x$.

14. $y = \log \sec x$, $\frac{dy}{dx} = \tan x$.

Employing the rule of § 14 for the differentiation of a function of a function,

$$\frac{dy}{dx} = \frac{\sec x \tan x}{\sec x} = \tan x.$$

15. $y = \log \sin x$, $\frac{dy}{dx} = \cot x$.

16. $y = \tan x - x$, $\frac{dy}{dx} = \tan^2 x$.

17. $y = \cot x + x$, $\frac{dy}{dx} = -\cot^2 x$.

18. $y = \frac{1}{2}\tan^2 x + \log \cos x$, $\frac{dy}{dx} = \tan^3 x$.

19. $y = \frac{1}{3}\tan^3 x - \tan x + x$, $\frac{dy}{dx} = \tan^4 x$.

20. $y = \log \tan \frac{1}{2}x$, $\frac{dy}{dx} = \operatorname{cosec} x$.

By § 14,

$$\frac{dy}{dx} = \frac{\sec^2 \frac{1}{2}x \times \frac{1}{2}}{\tan \frac{1}{2}x} = \frac{1}{2 \sin \frac{1}{2}x \cos \frac{1}{2}x} = \frac{1}{\sin x} = \operatorname{cosec} x.$$

21. $y = \log \tan(\frac{1}{4}\pi + \frac{1}{2}x), \dfrac{dy}{dx} = \sec x.$

By § 14,

$$\frac{dy}{dx} = \frac{\sec^2(\frac{1}{4}\pi + \frac{1}{2}x) \times \frac{1}{2}}{\tan(\frac{1}{4}\pi + \frac{1}{2}x)} = \frac{1}{2\sin(\frac{1}{4}\pi + \frac{1}{2}x)\cos(\frac{1}{4}\pi + \frac{1}{2}x)}$$

$$= \frac{1}{\sin(\frac{1}{2}\pi + x)} = \frac{1}{\cos x}$$

22. $y = \log\sqrt{\left(\dfrac{1+\sin x}{1-\sin x}\right)}, \dfrac{dy}{dx} = \sec x.$

Here

$$y = \tfrac{1}{2}\log(1+\sin x) - \tfrac{1}{2}\log(1-\sin x),$$

$$\frac{dy}{dx} = \tfrac{1}{2}\frac{\cos x}{1+\sin x} - \tfrac{1}{2}\frac{-\cos x}{1-\sin x}$$

$$= \frac{\cos x}{1-\sin^2 x} = \frac{1}{\cos x} = \sec x.$$

23. $y = \sec x \tan x + \log \tan(\frac{1}{4}\pi + \frac{1}{2}x), \dfrac{dy}{dx} = 2\sec^3 x.$

24. $y = \tfrac{1}{3}\tan^3 x + \tan x, \dfrac{dy}{dx} = \sec^4 x.$

25. $y = \sin^{-1}2x\sqrt{(1-x^2)}, \dfrac{dy}{dx} = \dfrac{2}{\sqrt{(1-x^2)}}.$

By § 14,

$$\frac{dy}{dx} = \frac{2\sqrt{(1-x^2)} - \dfrac{2x^2}{\sqrt{(1-x^2)}}}{\sqrt{\{1-4x^2(1-x^2)\}}} = \frac{2}{\sqrt{(1-x^2)}}.$$

26. $y = \sin^{-1}\dfrac{2x}{1+x^2}, \dfrac{dy}{dx} = \dfrac{2}{1+x^2}$

By § 14,

$$\frac{dy}{dx} = \frac{\dfrac{2(1+x^2) - 4x^2}{(1+x^2)^2}}{\sqrt{\left\{1 - \dfrac{4x^2}{(1+x^2)^2}\right\}}} = \frac{2}{1+x^2}.$$

27. $y = \cos^{-1}(1-2x^2), \dfrac{dy}{dx} = \dfrac{2}{\sqrt{(1-x^2)}}.$

28. $y = \sin^{-1}(3x - 4x^3)$, $\dfrac{dy}{dx} = \dfrac{3}{\sqrt{(1-x^2)}}$.

29. $y = \sin^{-1}4x(1 - 2x^2)\sqrt{(1-x^2)}$, $\dfrac{dy}{dx} = \dfrac{4}{\sqrt{(1-x^2)}}$

30. $y = \tan^{-1}\dfrac{2x}{1-x^2}$, $\dfrac{dy}{dx} = \dfrac{2}{1+x^2}$.

31. $y = \tan^{-1}\dfrac{3x - x^3}{1 - 3x^2}$, $\dfrac{dy}{dx} = \dfrac{3}{1+x^2}$.

32. $y = \tan^{-1}\dfrac{ax+b}{\sqrt{(ac-b^2)}}$, $\dfrac{dy}{dx} = \dfrac{\sqrt{(ac-b^2)}}{ax^2+2bx+c}$.

Then

$$\frac{dy}{dx} = \frac{\dfrac{a}{\sqrt{(ac-b^2)}}}{1+\dfrac{(ax+b)^2}{ac-b^2}} = \frac{\sqrt{(ac-b^2)}}{ax^2+2bx+c}.$$

33. $y = \tan^{-1}\dfrac{x\sqrt{(bp-aq)}}{\sqrt{\{a(p+qx^2)\}}} = \cos^{-1}\sqrt{\left\{\dfrac{a(p+qx^2)}{p(a+bx^2)}\right\}}$,

$$\frac{dy}{dx} = \frac{\sqrt{\{a(bp-aq)\}}}{(a+bx^2)\sqrt{(p+qx^2)}}.$$

34. $y = \sec^{-1}\dfrac{x^2}{2-x^2}$, $\dfrac{dy}{dx} = \dfrac{2}{x\sqrt{(x^2-1)}}$.

35. $y = \mathrm{vers}^{-1}(4x - 2x^2)$, $\dfrac{dy}{dx} = \dfrac{2}{\sqrt{(2x-x^2)}}$.

36. $y = \log\sqrt{\left(\dfrac{1+x}{1-x}\right)}$, $\dfrac{dy}{dx} = \dfrac{1}{1-x^2}$.

37. $y = \log\sqrt{\left(\dfrac{x-1}{x+1}\right)}$, $\dfrac{dy}{dx} = \dfrac{1}{x^2-1}$.

38. $y = \tfrac{1}{6}\log\dfrac{(x+1)^3}{x^3+1} + \dfrac{1}{\sqrt{3}}\tan^{-1}\dfrac{2x-1}{\sqrt{3}}$, $\dfrac{dy}{dx} = \dfrac{1}{x^3+1}$.

Here

$$y = \tfrac{1}{2}\log(x+1) - \tfrac{1}{6}\log(x^3+1) + \frac{1}{\sqrt{3}}\tan^{-1}\frac{2x-1}{\sqrt{3}},$$

$$\frac{dy}{dx} = \tfrac{1}{2}\frac{1}{x+1} - \tfrac{1}{6}\frac{3x^2}{x^3+1} + \frac{1}{\sqrt{3}}\frac{\frac{2}{\sqrt{3}}}{1+\frac{(2x-1)^2}{3}}$$

$$= \tfrac{1}{2}\frac{1}{x+1} - \tfrac{1}{2}\frac{x^2}{x^3+1} + \tfrac{1}{2}\frac{1}{x^2-x+1}$$

$$= \tfrac{1}{2}\frac{x^2-x+1-x^2+x+1}{x^3+1} = \frac{1}{x^3+1}.$$

39. $y = \log\sqrt[4]{\left(\frac{x-1}{x+1}\right)} - \tfrac{1}{2}\tan^{-1}x,\ \dfrac{dy}{dx} = \dfrac{1}{x^4-1}.$

40. $y = \log\sqrt{\left(\frac{x^2+x\sqrt{2}+1}{x^2-x\sqrt{2}+1}\right)} + \tan^{-1}\dfrac{x\sqrt{2}}{1-x^2},\ \dfrac{dy}{dx} = \dfrac{2\sqrt{2}}{x^4+1}.$

41. $y = \log\{x + \sqrt{(x^2-a^2)}\},\ \dfrac{dy}{dx} = \dfrac{1}{\sqrt{(x^2-a^2)}}.$

42. $y = \log\dfrac{\sqrt{(a^2+x^2)}-a}{x},\ \dfrac{dy}{dx} = \dfrac{a}{x\sqrt{(a^2+x^2)}}.$

43. $y = \log\sqrt{\left[\dfrac{\sqrt{\{a(p+qx^2)\}}+x\sqrt{(aq-bp)}}{\sqrt{\{a(p+qx^2)\}}-x\sqrt{(aq-bp)}}\right]},$

$$\dfrac{dy}{dx} = \dfrac{\sqrt{\{a(aq-bp)\}}}{(a+bx^2)\sqrt{(p+qx^2)}}.$$

√ 44. $y = \tfrac{1}{4}\sinh 2x - \tfrac{1}{2}x,\ \dfrac{dy}{dx} = \sinh^2 x.$

‵ 45. $y = \tfrac{1}{4}\sinh 2x + \tfrac{1}{2}x,\ \dfrac{dy}{dx} = \cosh^2 x.$

46. $y = \tfrac{1}{3}\cosh^3 x - \cosh x,\ \dfrac{dy}{dx} = \sinh^3 x.$

47. $y = \tfrac{1}{32}\sinh 4x + \tfrac{1}{4}\sinh 2x + \tfrac{3}{8}x,\ \dfrac{dy}{dx} = \cosh^4 x.$

48. $y = \log\cosh x,\ \dfrac{dy}{dx} = \tanh x.$

49. $y = \log\sinh x,\ \dfrac{dy}{dx} = \coth x.$

50. $y = x - \tanh x,\ \dfrac{dy}{dx} = \tanh^2 x.$

51. $y = \log \cosh x - \tfrac{1}{2}\tanh^2 x,\ \dfrac{dy}{dx} = \tanh^3 x.$

52. $y = x - \tanh x - \tfrac{1}{3}\tanh^3 x,\ \dfrac{dy}{dx} = \tanh^4 x.$

53. $y = 2\tan^{-1}\tanh \tfrac{1}{2}x,\ \dfrac{dy}{dx} = \operatorname{sech} x.$

54. $y = \log \tanh \tfrac{1}{2}x,\ \dfrac{dy}{dx} = \operatorname{cosech} x.$

55. $y = \tfrac{1}{2}\operatorname{cosech} x \coth x + \tfrac{1}{2}\log \tanh \tfrac{1}{2}x,\ \dfrac{dy}{dx} = \operatorname{cosech}^3 x.$

56. $y = \tanh x - \tfrac{1}{3}\tanh^3 x,\ \dfrac{dy}{dx} = \operatorname{sech}^4 x.$

57. $y = \sinh^{-1}2x\sqrt{(x^2-1)},\ \dfrac{dy}{dx} = \dfrac{2}{\sqrt{(x^2-1)}}.$

58. $y = \sinh^{-1}\dfrac{2x}{1-x^2},\ \dfrac{dy}{dx} = \dfrac{2}{1-x^2}.$

59. $y = \cosh^{-1}(2x^2-1),\ \dfrac{dy}{dx} = \dfrac{2}{\sqrt{(x^2-1)}}.$

60. $y = \sinh^{-1}(3x+4x^3),\ \dfrac{dy}{dx} = \dfrac{3}{\sqrt{(1+x^2)}}.$

61. $y = \tanh^{-1}\dfrac{ax+b}{\sqrt{(b^2-ac)}},\ \dfrac{dy}{dx} = \dfrac{\sqrt{(b^2-ac)}}{ax^2+2bx+c}.$

62. $y = \tanh^{-1}\dfrac{3x+x^3}{1+3x^2},\ \dfrac{dy}{dx} = \dfrac{3}{1-x^2}.$

63. $y = \tan^{-1}x + \tanh^{-1}x,\ \dfrac{dy}{dx} = \dfrac{2}{1-x^4}.$

64. $y = \tan^{-1}\dfrac{x\sqrt{2}}{1-x^2} + \tanh^{-1}\dfrac{x\sqrt{2}}{1+x^2},\ \dfrac{dy}{dx} = \dfrac{2\sqrt{2}}{1+x^4}.$

65. $y = \tanh^{-1}\dfrac{x\sqrt{(aq-bp)}}{\sqrt{\{a(p+qx^2)\}}} = \cosh^{-1}\sqrt{\left\{\dfrac{a(p+qx^2)}{p(a+bx^2)}\right\}}$,

$$\frac{dy}{dx} = \frac{\sqrt{\{a(aq-bp)\}}}{(a+bx^2)\sqrt{(p+qx^2)}}.$$

66. $y = e^{ax}\cos bx, \dfrac{dy}{dx} = ae^{ax}\cos(bx+a)\sec a,$ where $\tan a = \dfrac{b}{a}$.

By the rule of § 34 for the differentiation of a product

$$\frac{dy}{dx} = ae^{ax}\cos bx - e^{ax}b\sin bx$$
$$= ae^{ax}(\cos bx - \tan a \sin bx)$$
$$= ae^{ax}\cos(bx+a)\sec a.$$

67. $y = m\sin mx \cosh nx + n\cos mx \sinh nx,$

$$\frac{dy}{dx} = (m^2+n^2)\cos mx \cosh nx.$$

68. $y = x\sqrt{(a^2-x^2)} + a^2\sin^{-1}\dfrac{x}{a}, \dfrac{dy}{dx} = 2\sqrt{(a^2-x^2)}.$

69. $y = a^2\mathrm{vers}^{-1}\dfrac{x}{a} - (a-x)\sqrt{(2ax-x^2)},$

$$\frac{dy}{dx} = 2\sqrt{(2ax-x^2)}.$$

70. $y = x\sqrt{(x^2-a^2)} - a^2\cosh^{-1}\dfrac{x}{a}, \dfrac{dy}{dx} = 2\sqrt{(x^2-a^2)}.$

71. $y = x\sqrt{(a^2+x^2)} + a^2\sinh^{-1}\dfrac{x}{a}, \dfrac{dy}{dx} = 2\sqrt{(a^2+x^2)}.$

72. $y = x\sqrt{(2ax+x^2)} - a^2\cosh^{-1}\left(1+\dfrac{x}{a}\right),$

$$\frac{dy}{dx} = 2\sqrt{(2ax+x^2)}.$$

73. $y = \sin^{-1}\sqrt{\dfrac{x-\beta}{a-\beta}} = \cos^{-1}\sqrt{\dfrac{a-x}{a-\beta}},$

$$\frac{dy}{dx} = \tfrac{1}{2}\frac{1}{\sqrt{(a-x\cdot x-\beta)}}.$$

74. $y = \sinh^{-1}\sqrt{\dfrac{x-a}{a-\beta}} = \cosh^{-1}\sqrt{\dfrac{x-\beta}{a-\beta}}$,

$$\frac{dy}{dx} = \tfrac{1}{2}\frac{1}{\sqrt{(x-a \cdot x-\beta)}}.$$

75. $y = \log\{\sqrt{(x-a)} + \sqrt{(x-\beta)}\}, \dfrac{dy}{dx} = \tfrac{1}{2}\dfrac{1}{\sqrt{(x-a \cdot x-\beta)}}$.

76. $y = \sin^{-1}\sqrt{(\sin 2x)} + \sinh^{-1}\sqrt{(\sin 2x)}, \dfrac{dy}{dx} = \sqrt{(2\cot x)}$.

77. $y = \tan^{-1}\sqrt{(\tanh x)} + \tanh^{-1}\sqrt{(\tanh x)}$,

$$\frac{dy}{dx} = \sqrt{(\coth x)}.$$

78. $y = \log\dfrac{\sqrt{(a+b)}\cos\tfrac{1}{2}x + \sqrt{(a-b)}\sin\tfrac{1}{2}x}{\sqrt{(a+b)}\cos\tfrac{1}{2}x - \sqrt{(a-b)}\sin\tfrac{1}{2}x}$,

$$= 2\tanh^{-1}\left\{ \left(\frac{a-b}{a+b}\right)^{\frac{1}{2}}\tan\tfrac{1}{2}x \right\}, \frac{dy}{dx} = \frac{\sqrt{(a^2-b^2)}}{a\cos x + b}.$$

79. $y = \tan^{-1}\left\{ \left(\dfrac{a-b}{a+b}\right)^{\frac{1}{2}}\tan\tfrac{1}{2}x \right\}, \dfrac{dy}{dx} = \tfrac{1}{2}\dfrac{\sqrt{(a^2-b^2)}}{a+b\cos x}$.

80. $y = \tanh^{-1}\left\{ \left(\dfrac{a-b}{a+b}\right)^{\frac{1}{2}}\tanh\tfrac{1}{2}x \right\}, \dfrac{dy}{dx} = \tfrac{1}{2}\dfrac{\sqrt{(a^2-b^2)}}{a+b\cosh x}$.

81. $y = \tan^{-1}\left\{ \left(\dfrac{a-b}{a+b}\right)^{\frac{1}{2}}\tanh\tfrac{1}{2}x \right\}, \dfrac{dy}{dx} = \tfrac{1}{2}\dfrac{\sqrt{(a^2-b^2)}}{a\cosh x + b}$.

82. $y = \cos^{-1}\dfrac{a\cos x + b}{a+b\cos x}, \dfrac{dy}{dx} = \dfrac{\sqrt{(a^2-b^2)}}{a+b\cos x}$.

By § 14,

$$\frac{dy}{dx} = -\frac{\dfrac{-a\sin x(a+b\cos x) + (a\cos x + b)b\sin x}{(a+b\cos x)^2}}{\sqrt{\left\{1 - \left(\dfrac{a\cos x + b}{a+b\cos x}\right)^2\right\}}}$$

$$= \frac{(a^2-b^2)\sin x}{(a+b\cos x)\sqrt{\{(a+b\cos x)^2 - (a\cos x + b)^2\}}} = \frac{\sqrt{(a^2-b^2)}}{a+b\cos x}.$$

83. $y = \cosh^{-1}\dfrac{a+b\cos x}{a\cos x + b}, \dfrac{dy}{dx} = \dfrac{\sqrt{(a^2-b^2)}}{a\cos x + b}$.

84. $y = \sinh^{-1}\dfrac{a \sinh x - b}{a + b \sinh x}$, $\dfrac{dy}{dx} = \dfrac{\sqrt{(a^2 + b^2)}}{a + b \sinh x}$.

85. $y = \cos^{-1}\dfrac{a(b \cos x + c \sin x) + b^2 + c^2}{(a + b \cos x + c \sin x)\sqrt{(b^2 + c^2)}}$,

$\dfrac{dy}{dx} = \dfrac{\sqrt{(a^2 - b^2 - c^2)}}{a + b \cos x + c \sin x}$.

86. $y = \cosh^{-1}\dfrac{a(b \cos x + c \sin x) + b^2 + c^2}{(a + b \cos x + c \sin x)\sqrt{(b^2 + c^2)}}$,

$\dfrac{dy}{dx} = \dfrac{\sqrt{(-a^2 + b^2 + c^2)}}{a + b \cos x + c \sin x}$.

87. $y = \cos^{-1}\dfrac{a(b \cosh x + c \sinh x) + b^2 - c^2}{(a + b \cosh x + c \sinh x)\sqrt{(b^2 - c^2)}}$,

$\dfrac{dy}{dx} = \dfrac{\sqrt{(-a^2 + b^2 - c^2)}}{a + b \cosh x + c \sinh x}$.

88. $y = \cosh^{-1}\dfrac{a(b \cosh x + c \sinh x) + b^2 - c^2}{(a + b \cosh x + c \sinh x)\sqrt{(b^2 - c^2)}}$,

$\dfrac{dy}{dx} = \dfrac{\sqrt{(a^2 - b^2 + c^2)}}{a + b \cosh x + c \sinh x}$.

89. $y = \sinh^{-1}\dfrac{a(b \cosh x + c \sinh x) + b^2 - c^2}{(a + b \cosh x + c \sinh x)\sqrt{(c^2 - b^2)}}$,

$\dfrac{dy}{dx} = \dfrac{\sqrt{(a^2 - b^2 + c^2)}}{a + b \cosh x + c \sinh x}$.

90. Prove that in the curve $y = a \log \sec \dfrac{x}{a}$,

$$\frac{ds}{dx} = \sec \frac{x}{a}, \text{ and } x = a\psi.$$

For $\quad \dfrac{dy}{dx} = \tan\dfrac{x}{a}$, and therefore $\dfrac{ds}{dx} = \sec\dfrac{x}{a}$;

and $\quad \tan\psi = \dfrac{dy}{dx} = \tan\dfrac{x}{a}$, so that $\psi = \dfrac{x}{a}$.

91. In the curve $y = a \cosh \dfrac{x}{a}$ (the *catenary*) prove that the length of MQ, the perpendicular drawn from M, the foot of the ordinate, on the tangent, is always equal to a.

92. Find the equation of the locus of Q (the *tractrix*), and prove that QM is the tangent at Q, and consequently of constant length.

93. Prove that the equation (§ 30)

$\theta = \text{gd } n\phi$, or $\tan \tfrac{1}{2}\theta = \tanh \tfrac{1}{2}n\phi$, or $\sin \theta = \tanh n\phi$,

connecting θ the latitude and ϕ the longitude, represents a curve on a terrestrial sphere cutting the meridians at a constant angle a, where $n = \cot a$ (the *loxodrome* or *rhumb line*).

94. Resolve $\sin x$ and $\cos x$ into factors and thence deduce by logarithmic differentiation that

$$\cot x = \frac{1}{x} + \frac{1}{x-\pi} + \frac{1}{x-2\pi} + \cdots$$

$$+ \frac{1}{x+\pi} + \frac{1}{x+2\pi} + \cdots,$$

$$\operatorname{cosec} x = \frac{1}{x} - \frac{1}{x-\pi} + \frac{1}{x-2\pi} - \cdots$$

$$- \frac{1}{x+\pi} + \frac{1}{x+2\pi} - \cdots;$$

and find the corresponding expressions for $\tan x$ and $\sec x$; also of $\sec^2 x$ and $\operatorname{cosec}^2 x$.

CHAPTER II.

INTEGRATION.

36. The process of *Integration* is the reverse of Differentiation, and is the province of the *Integral Calculus*.

In the Differential Calculus a function fx is given, and we investigate the rules for finding $\dfrac{dfx}{dx}$ or $f'x$.

But in the Integral Calculus the function $f'x$ is given, and we are required to find fx, the function of which $f'x$ is the differential coefficient.

This process is of a tentative nature, depending on a previous knowledge of differentiation as explained in Chapter I.; just as Division in Arithmetic is a tentative process, depending on a knowledge of Multiplication.

With the notation of the Differential Calculus

$$f'x = \frac{dfx}{dx},$$

or $\qquad\qquad f'x\,dx = dfx,$

the notation of *Differentials*.

Now supposing f and d to represent inverse operations, so that f and d cancel; operating by f,

$$\int f'x\,dx = \int dfx,$$
$$= fx,$$

the notation of the Integral Calculus.

37. To every differentiation in the Differential Calculus corresponds an integration in the Integral Calculus, and this correspondence for the functions we shall employ is exhibited in the following table.

DIFFERENTIAL CALCULUS.	INTEGRAL CALCULUS.
(a) $\dfrac{dx^n}{dx} = nx^{n-1}.$	$\int x^m dx = \dfrac{x^{m+1}}{m+1}.$
(b) $\dfrac{d \sin x}{dx} = \cos x.$	$\int \cos x \, dx = \sin x.$
(c) $\dfrac{d \cos x}{dx} = -\sin x.$	$\int \sin x \, dx = -\cos x.$
(d) $\dfrac{d \tan x}{dx} = \sec^2 x.$	$\int \sec^2 x \, dx = \tan x.$
(e) $\dfrac{d \cot x}{dx} = -\operatorname{cosec}^2 x.$	$\int \operatorname{cosec}^2 x \, dx = -\cot x.$
(f) $\dfrac{d \sec x}{dx} = \sec x \tan x.$	$\int \sec x \tan x \, dx = \sec x.$
(g) $\dfrac{d \operatorname{cosec} x}{dx} = -\operatorname{cosec} x \cot x.$	$\int \operatorname{cosec} x \cot x \, dx = -\operatorname{cosec} x.$
(h) $\dfrac{d \operatorname{vers} x}{dx} = \sin x.$	$\int \sin x \, dx = \operatorname{vers} x.$
(i) $\dfrac{d \sin^{-1}\frac{x}{a}}{dx} = \dfrac{1}{\sqrt{(a^2-x^2)}}.$	$\int \dfrac{dx}{\sqrt{(a^2-x^2)}} = \sin^{-1}\dfrac{x}{a}.$
(j) $\dfrac{d \cos^{-1}\frac{x}{a}}{dx} = -\dfrac{1}{\sqrt{(a^2-x^2)}}.$	$\int \dfrac{dx}{\sqrt{(a^2-x^2)}} = -\cos^{-1}\dfrac{x}{a}.$
(k) $\dfrac{d \tan^{-1}\frac{x}{a}}{dx} = \dfrac{a}{a^2+x^2}.$	$\int \dfrac{dx}{a^2+x^2} = \dfrac{1}{a}\tan^{-1}\dfrac{x}{a}.$
(l) $\dfrac{d \cot^{-1}\frac{x}{a}}{dx} = -\dfrac{a}{x^2+a^2}.$	$\int \dfrac{dx}{x^2+a^2} = -\dfrac{1}{a}\cot^{-1}\dfrac{x}{a}.$
(m) $\dfrac{d \sec^{-1}\frac{x}{a}}{dx} = \dfrac{a}{x\sqrt{(x^2-a^2)}}.$	$\int \dfrac{dx}{x\sqrt{(x^2-a^2)}} = \dfrac{1}{a}\sec^{-1}\dfrac{x}{a}.$

DIFFERENTIAL CALCULUS.	INTEGRAL CALCULUS.
(n) $\dfrac{d \operatorname{cosec}^{-1}\frac{x}{a}}{dx} = -\dfrac{a}{x\sqrt{(x^2-a^2)}}.$	$\displaystyle\int \dfrac{dx}{x\sqrt{(x^2-a^2)}} = -\dfrac{1}{a}\operatorname{cosec}^{-1}\dfrac{x}{a}.$
(o) $\dfrac{d \operatorname{vers}^{-1}\frac{x}{a}}{dx} = \dfrac{1}{\sqrt{(2ax-x^2)}}.$	$\displaystyle\int \dfrac{dx}{\sqrt{(2ax-x^2)}} = \operatorname{vers}^{-1}\dfrac{x}{a}.$
(p) $\dfrac{da^x}{dx} = a^x \log a.$	$\displaystyle\int a^x dx = \dfrac{a^x}{\log a}.$
(q) $\dfrac{de^{cx}}{dx} = ce^{cx}.$	$\displaystyle\int e^{cx}dx = \dfrac{e^{cx}}{c}.$
(r) $\dfrac{d \sinh x}{dx} = \cosh x.$	$\displaystyle\int \cosh x\, dx = \sinh x.$
(s) $\dfrac{d \cosh x}{dx} = \sinh x.$	$\displaystyle\int \sinh x\, dx = \cosh x.$
(t) $\dfrac{d \tanh x}{dx} = \operatorname{sech}^2 x.$	$\displaystyle\int \operatorname{sech}^2 x\, dx = \tanh x.$
(u) $\dfrac{d \coth x}{dx} = -\operatorname{cosech}^2 x.$	$\displaystyle\int \operatorname{cosech}^2 x\, dx = -\coth x.$
(v) $\dfrac{d \log x}{dx} = \dfrac{1}{x}.$	$\displaystyle\int \dfrac{dx}{x} = \log x.$
(w) $\dfrac{d \sinh^{-1}\frac{x}{a}}{dx} = \dfrac{1}{\sqrt{(a^2+x^2)}}.$	$\displaystyle\int \dfrac{dx}{\sqrt{(a^2+x^2)}} = \sinh^{-1}\dfrac{x}{a},$ or $= \log\left\{ \dfrac{\sqrt{(a^2+x^2)}+x}{a} \right\}.$
(x) $\dfrac{d \cosh^{-1}\frac{x}{a}}{dx} = \dfrac{1}{\sqrt{(x^2-a^2)}}.$	$\displaystyle\int \dfrac{dx}{\sqrt{(x^2-a^2)}} = \cosh^{-1}\dfrac{x}{a},$ or $= \log\left\{ x + \sqrt{(x^2-a^2)} \right\}.$
(y) $\dfrac{d \tanh^{-1}\frac{x}{a}}{dx} = \dfrac{a}{a^2-x^2}.$	$\displaystyle\int \dfrac{dx}{a^2-x^2} = \dfrac{1}{a}\tanh^{-1}\dfrac{x}{a}\,(x<a),$ or $= \dfrac{1}{2a}\log\dfrac{a+x}{a-x}.$
(z) $\dfrac{d \coth^{-1}\frac{x}{a}}{dx} = -\dfrac{a}{x^2-a^2}.$	$\displaystyle\int \dfrac{dx}{x^2-a^2} = -\dfrac{1}{a}\coth^{-1}\dfrac{x}{a}\,(x>a)$ or $= \dfrac{1}{2a}\log\dfrac{x-a}{x+a}.$

Since the d.c. of a constant is zero, therefore in integration an arbitrary constant may be added, and the integral is then called an *indefinite* integral.

When, in the above formulæ, a discrepancy of results of integration appears, as between (c) and (h), (i) and (j), (k) and (l), (m) and (n), the results will be found to differ by a constant.

38. Any function of x which can be thrown into the form $(fx)^m f'x$ is immediately integrable by (a), because

$$\int (fx)^m f'x\, dx = \frac{(fx)^{m+1}}{m+1}.$$

But if $m = -1$, then by (v)

$$\int \frac{f'x}{fx}\, dx = \log fx.$$

Examples.—Integrate with respect to x:

$$0, 1, a, x, ax+b, x^2, x^3, \sqrt{x}, \sqrt[3]{x}, \frac{1}{\sqrt{x}}, \frac{1}{\sqrt[3]{x}}, \frac{1}{x}, \frac{1}{x^{\frac{3}{2}}}, \frac{1}{x^{\frac{4}{3}}}, \frac{1}{x^{2}}, \frac{1}{x^{n}}, \frac{1}{(nx)^{\frac{n-1}{n}}},$$

$$x+a, \ (x+a)^2, \ \sqrt{(x+a)}, \ (x+a)^{\frac{2}{3}}, \ \frac{1}{\sqrt{(x+a)}}, \ \frac{1}{x+a}, \ \frac{1}{(x+a)^{\frac{3}{2}}},$$

$$(mx+n)^p, \quad ax^2 + 2bx + c, \quad (x^2 + a^2)^3, \quad \frac{B}{(x-b)^n}, \quad x\sqrt{(a^2 + x^2)},$$

$$\frac{x}{\sqrt{(x^2 - a^2)}}, \quad \frac{x}{(a^2 - x^2)^{\frac{5}{2}}}, \quad \frac{x-c}{\sqrt{\{(x-c)^2 + a^2\}}}, \quad \frac{x-c}{a^2 - (x-c)^2},$$

$$\{(x-c)^2 - a^2\}^m (x-c), \quad \frac{ax+b}{\sqrt{(ax^2 + 2bx + c)}}, \quad \frac{ax+b}{ax^2 + 2bx + c},$$

$(ax^2+2bx+c)^m(ax+b)$, $\dfrac{1}{(x+a)^{\frac{1}{2}}+(x+b)^{\frac{1}{2}}}$ (rationalize the denominator).

(It is not necessary to give the answers, as the correctness of the result can always be tested by differentiation.)

39. *Integration of Rational Fractions.*

Any rational fraction $\dfrac{fx}{\mathrm{F}x}$, the numerator and denominator of which are rational integral algebraical functions of x, and therefore of the form

$$\frac{fx}{\mathrm{F}x} = \frac{ax^m + bx^{m-1} + cx^{m-2} + \ldots}{Ax^n + Bx^{n-1} + Cx^{n-2} + \ldots},$$

where m and n are positive integers, is integrated by resolving it into its partial functions by the ordinary rules of Algebra; if the degree of the numerator is equal to or greater than the degree of the denominator, that is, if $m = n$ or $> n$, the quotient must be first obtained by division.

Example 1. To integrate $\dfrac{x^3}{x^2 - 3x + 2}$, we must suppose

it resolved into $x + 3 + \dfrac{A}{x-1} + \dfrac{B}{x-2}$, $x+3$ being the

quotient, and $\dfrac{A}{x-1}$, $\dfrac{B}{x-2}$, the partial fractions.

To determine A, multiply by its denominator $x-1$;

then $$\frac{x^3}{x-2}=(x+3)(x-1)+A+B\frac{x-1}{x-2}.$$

Now, put $x-1=0$; then $A=\dfrac{x^3}{x-2}$ (when $x-1=0$)$=-1$.

Similarly $\quad B=\dfrac{x^3}{x-1}$ (when $x-2=0$)$=8$.

Then $$\int\frac{x^3dx}{x^2-3x+2}$$

$$=\int\left(x+3-\frac{1}{x-1}+\frac{8}{x-2}\right)dx$$

$$=\tfrac{1}{2}x^2+3x-\log(x-1)+8\log(x-2).$$

Generally, in integrating $\dfrac{fx}{Fx}$, if $x-a$ is a factor of the denominator Fx, so that

$$Fx=(x-a)\phi x,$$

to determine the corresponding partial fraction $\dfrac{A}{x-a}$,

assume $$\frac{fx}{Fx}=\frac{A}{x-a}+\frac{R}{\phi x};$$

then $$\frac{fx}{\phi x}=A+\frac{R}{\phi x}(x-a).$$

Put $x-a=0$, then

$$A=\frac{fa}{\phi a}.$$

But $$F'x=\phi x+(x-a)\phi' x,$$
and therefore $\quad F'a=\phi a,$

so that $$A=\frac{fa}{F'a}.$$

Example 2. To integrate $\dfrac{x^{m-1}}{x^n-1}$; if $x-a$ denotes a factor of the denominator x^n-1, then $a=1$, or by De Moivre's Theorem,

$$a = \cos\frac{2r\pi}{n} + i\sin\frac{2r\pi}{n},$$

and r has the values $1, -1, 2, -2, \ldots$.

Let $\qquad \dfrac{x^{m-1}}{x^n-1} = \dfrac{A}{x-1} + \Sigma\dfrac{A_r}{x-a}$;

here $fa=a^{m-1}$, $Fa=a^n-1$, $F'a=na^{n-1}$;

so that $\qquad A_r = \dfrac{1}{n}\dfrac{a^{m-1}}{a^{n-1}} = \dfrac{1}{n}a^m$ (since $a^n=1$)

$$= \frac{1}{n}\left(\cos\frac{2mr\pi}{n} + i\sin\frac{2mr\pi}{n}\right)$$

by De Moivre's Theorem; also

$$A = \frac{1}{n}.$$

Therefore

$$\int\frac{x^{m-1}dx}{x^n-1} = \frac{1}{n}\log(x-1)$$

$$+ \frac{1}{n}\Sigma\left(\cos\frac{2mr\pi}{n} + i\sin\frac{2mr\pi}{n}\right)\log\left(x-\cos\frac{2r\pi}{n} - i\sin\frac{2r\pi}{n}\right).$$

To express this result in a real form, the partial fractions with numerators A_r and A_{-r}, corresponding to conjugate imaginary n^{th} roots of unity must be combined into a real form as follows:

$$\frac{A_r}{x-\cos\dfrac{2r\pi}{n} - i\sin\dfrac{2r\pi}{n}} + \frac{A_{-r}}{x-\cos\dfrac{2r\pi}{n} + i\sin\dfrac{2r\pi}{n}}$$

$$= \frac{1}{n} \cos \frac{2mr\pi}{n} \cdot \frac{2x - 2\cos\dfrac{2r\pi}{n}}{x^2 - 2x\cos\dfrac{2r\pi}{n} + 1}$$

$$- \frac{2}{n} \sin \frac{2mr\pi}{n} \cdot \frac{\sin\dfrac{2r\pi}{n}}{x^2 - 2x\cos\dfrac{2r\pi}{n} + 1} \; ;$$

and the corresponding integrals are

$$\frac{1}{n} \cos \frac{2mr\pi}{n} \log \left(x^2 - 2x\cos\frac{2r\pi}{n} + 1 \right)$$

$$- \frac{2}{n} \sin \frac{2mr\pi}{n} \tan^{-1} \frac{x - \cos\dfrac{2r\pi}{n}}{\sin\dfrac{2r\pi}{n}}.$$

40. In general, when Fx has a pair of conjugate imaginary factors, $x - a - i\beta$ and $x - a + i\beta$, combining into a real quadratic product $(x-a)^2 + \beta^2$, it is usual to assume a corresponding partial fraction of the form

$$\frac{Px+Q}{(x-a)^2+\beta^2},$$

obtained by combining the partial fractions

$$\frac{A+iB}{x-a-i\beta} \text{ and } \frac{A-iB}{x-a+i\beta}.$$

Then
$$\frac{fx}{Fx} = \frac{Px+Q}{(x-a)^2+\beta^2} + \frac{R}{\phi x},$$

suppose; or

$$fx = (Px+Q)\phi x + R\{(x-a)^2+\beta^2\}.$$

To determine P and Q, put $(x-a)^2+\beta^2=0$; this gives imaginary values to x, namely, $a+i\beta$ and $a-i\beta$; but by continually substituting $2ax - a^2 - \beta^2$ for x^2, f_x can be reduced to the form $Lx + M$, and $(Px + Q)\phi x$ to the form $L'x + M'$; and then $L = L'$, $M = M'$, whence P and Q are determined.

Example 3. To integrate $\dfrac{1}{x^4 + 1}$; here the denominator $x^4 + 1$ splits up into two quadratic factors $x^2 + x\sqrt{2} + 1$ and $x^2 - x\sqrt{2} + 1$; therefore assume

$$\frac{1}{x^4 + 1} = \frac{Px + Q}{x^2 + x\sqrt{2} + 1} + \frac{Rx + S}{x^2 - x\sqrt{2} + 1};$$

then

$$1 = (Px + Q)(x^2 - x\sqrt{2} + 1) + (Rx + S)(x^2 + x\sqrt{2} + 1).$$

Put $x^2 + x\sqrt{2} + 1 = 0$, or $x^2 = -x\sqrt{2} - 1$; then

$$1 = (Px + Q)(-2x\sqrt{2})$$
$$= -2Px^2\sqrt{2} - 2Qx\sqrt{2}$$
$$= 2P\sqrt{2}(x\sqrt{2} + 1) - 2Qx\sqrt{2}$$
$$= (4P - 2Q\sqrt{2})x + 2P\sqrt{2},$$

of the form $Lx + M$, and therefore

$$4P - 2Q\sqrt{2} = 0, \qquad 2P\sqrt{2} = 1,$$

or $$P = \tfrac{1}{4}\sqrt{2}, \; Q = \tfrac{1}{2}.$$

Again, put $x^2 - x\sqrt{2} + 1 = 0$; then

$$1 = (Rx + S)2x\sqrt{2}$$
$$= 2R\sqrt{2}(x\sqrt{2} - 1) + 2Sx\sqrt{2};$$

and therefore

$$4R + 2S\sqrt{2} = 0, \qquad 2R\sqrt{2} = -1;$$

or $$R = -\tfrac{1}{4}\sqrt{2}, \; S = \tfrac{1}{2}.$$

Therefore
$$\int \frac{dx}{x^4+1}$$

$$= \tfrac{1}{4}\sqrt{2}\int \frac{x+\sqrt{2}}{x^2+x\sqrt{2}+1}dx - \tfrac{1}{4}\sqrt{2}\int \frac{x-\sqrt{2}}{x^2-x\sqrt{2}+1}dx$$

$$= \tfrac{1}{8}\sqrt{2}\int \frac{2x+\sqrt{2}}{x^2+x\sqrt{2}+1}dx - \tfrac{1}{8}\sqrt{2}\int \frac{2x-\sqrt{2}}{x^2-x\sqrt{2}+1}dx$$

$$+ \tfrac{1}{4}\int \frac{dx}{x^2+x\sqrt{2}+1} + \tfrac{1}{4}\int \frac{dx}{x^2-x\sqrt{2}+1}$$

$$= \tfrac{1}{8}\sqrt{2}\log(x^2+x\sqrt{2}+1) - \tfrac{1}{8}\sqrt{2}\log(x^2-x\sqrt{2}+1)$$
$$+ \tfrac{1}{4}\sqrt{2}\tan^{-1}(x\sqrt{2}+1) + \tfrac{1}{4}\sqrt{2}\tan^{-1}(x\sqrt{2}-1)$$

$$= \tfrac{1}{8}\sqrt{2}\log\frac{x^2+x\sqrt{2}+1}{x^2-x\sqrt{2}+1} + \tfrac{1}{4}\sqrt{2}\tan^{-1}\frac{x\sqrt{2}}{1-x^2}$$

$$= \tfrac{1}{4}\sqrt{2}\left(\tanh^{-1}\frac{x\sqrt{2}}{1+x^2} + \tan^{-1}\frac{x\sqrt{2}}{1-x^2}\right).$$

41. When Fx has a factor $x-b$ repeated p times, so that $Fx = (x-b)^p\phi x$,
we must assume

$$\frac{fx}{Fx} = \frac{B_p}{(x-b)^p} + \frac{B_{p-1}}{(x-b)^{p-1}} + \ldots + \frac{B_1}{x-p} + \frac{R}{\phi x},$$

and then

$$\frac{fx}{\phi x} = B_p + B_{p-1}(x-b) + \ldots + B_1(x-b)^{p-1} + \frac{R}{\phi x}(x-b)^p.$$

Now put $x-b=y$, so that $x=b+y$, then

$$\frac{f(b+y)}{\phi(b+y)} = B_p + B_{p-1}y + \ldots + B_1 y^{p-1} + \frac{Ry^p}{\phi(b+y)},$$

so that $B_p, B_{p-1}, \ldots, B_1$ are the coefficients of $y^0, y^1, \ldots, y^{p-1}$,
in the expansion of $\dfrac{f(b+y)}{\phi(b+y)}$ in ascending powers of y.

Example 4. To integrate $\dfrac{(x^2+1)^2}{(x^2-1)^3}$, suppose it resolved into the partial fractions

$$\frac{B_3}{(x-1)^3}+\frac{B_2}{(x-1)^2}+\frac{B_1}{x-1}+\frac{C_3}{(x+1)^3}+\frac{C_2}{(x+1)^2}+\frac{C_1}{x+1}.$$

Then $B_3+B_2y+B_1y^2$ are the first three terms in the expansion of

$$\frac{(2+2y+y^2)^2}{(2+y)^3}=\tfrac{1}{2}(1+y+\tfrac{1}{2}y^2)^2(1+\tfrac{1}{2}y)^{-3}$$
$$=\tfrac{1}{2}+\tfrac{1}{4}y+\tfrac{1}{4}y^2\ldots$$

so that $\qquad B_3=\tfrac{1}{2},\quad B_2=\tfrac{1}{4},\quad B_1=\tfrac{1}{4}.$

Again $C_3+C_2y+C_1y^2$ are the first three terms in the expansion of

$$\frac{(2-2y+y^2)^2}{(-2+y)^3}=-\tfrac{1}{2}(1-y+\tfrac{1}{2}y^2)^2(1-\tfrac{1}{2}y)^{-3}$$
$$=-\tfrac{1}{2}+\tfrac{1}{4}y-\tfrac{1}{4}y^2\ldots$$

so that $\qquad C_3=-\tfrac{1}{2},\quad C_2=\tfrac{1}{4},\quad C_1=-\tfrac{1}{4}.$

Then $\displaystyle\int\frac{(x^2+1)^2}{(x^2-1)^3}dx$

$$=\int\Big\{\tfrac{1}{2}\frac{1}{(x-1)^3}+\tfrac{1}{4}\frac{1}{(x-1)^2}+\tfrac{1}{4}\frac{1}{x-1}$$
$$-\tfrac{1}{2}\frac{1}{(x+1)^3}+\tfrac{1}{4}\frac{1}{(x+1)^2}-\tfrac{1}{4}\frac{1}{x+1}\Big\}dx$$
$$=-\tfrac{1}{4}\frac{1}{(x-1)^2}-\tfrac{1}{4}\frac{1}{x-1}+\tfrac{1}{4}\log(x-1)$$
$$+\tfrac{1}{4}\frac{1}{(x+1)^2}-\tfrac{1}{4}\frac{1}{x+1}-\tfrac{1}{4}\log(x+1)$$
$$=-\tfrac{1}{2}x\frac{x^2+1}{(x^2-1)^2}+\tfrac{1}{4}\log\frac{x-1}{x+1}.$$

Any integral of the form

$$\int \frac{A + B(a + bx)^{\frac{p}{r}}}{C + D(a + bx)^{\frac{q}{r}}} dx,$$

where A, B, C, D are rational integral functions of x, can be made to depend on the integration of a rational fraction by the *substitution* of

$$a + bx = y^r.$$

Then　　　　　　$$bdx = ry^{r-1}dy,$$

and the integral

$$= \int \frac{A + By^p}{C + Dy^q} \frac{ry^{r-1}}{b} dy,$$

which is integrated by the preceding rules.

Examples.—Integrate with respect to x:

$$\frac{x^3}{x^2 - 1}, \quad \frac{1}{x^3 - 1}, \quad \frac{x}{x^3 + 1}, \quad \frac{x^5}{x^4 - 1}, \quad \frac{x^2}{x^3 - 5x^2 + 8x - 4}, \quad \frac{1}{x^4 - 1^r}$$

$$\frac{x}{x^4 + 1}, \quad \frac{x^2}{x^4 + 1}, \quad \frac{x^3}{x^4 + 1}, \quad \frac{x^2 - x + 2}{x^4 - 5x^2 + 4}, \quad \frac{1}{x^4 + 3x^2 + 2}, \quad \frac{x^3}{(1 - x)^2}$$

$$\frac{x^3}{(x + a)^4}, \quad \frac{Ax + B}{(mx + n)^p} \ (\text{put } mx + n = y), \quad \frac{1}{x^6 - 1}, \quad \frac{x^3}{x^9 + 1}, \quad \frac{1}{(x^3 + 1)^2}$$

$$\frac{1}{(x - a)^m(x - b)^n} \left(\text{put } \frac{x - a}{x - b} = y\right), \quad \frac{1}{(x + 1)\sqrt{(x + 2)}},$$

$$\frac{x}{(x + 2)\sqrt{(x + 1)}}, \quad x^2\sqrt{(x + a)}, \quad \frac{x^3}{\sqrt{(x - a)}}.$$

42. *Integration of circular and hyperbolic functions.*
To integrate powers and products of $\sin x$ and $\cos x$, the most general plan is to convert them into sines and cosines of multiples of x, which are immediately integrable.

Thus

$$\int \sin^2 x\, dx = \int \tfrac{1}{2}(1 - \cos 2x)\, dx = \tfrac{1}{2}x - \tfrac{1}{4}\sin 2x \, ;$$

$$\int \cos^2 x\, dx = \int \tfrac{1}{2}(1 + \cos 2x)\, dx = \tfrac{1}{2}x + \tfrac{1}{4}\sin 2x \, ;$$

$$\int \cos mx \cos nx\, dx = \int \{ \tfrac{1}{2}\cos(m-n)x + \tfrac{1}{2}\cos(m+n)x \}\, dx$$

$$= \tfrac{1}{2}\frac{\sin(m-n)x}{m-n} + \tfrac{1}{2}\frac{\sin(m+n)x}{m+n} \, ;$$

and similarly for $\int \sin mx \sin nx\, dx$ and $\int \sin mx \cos nx\, dx$.

To integrate $\sin^k x \cos^{2n+1} x$ and $\sin^{2n+1} x \cos^k x$, where n is an integer, write them in the form

$$\sin^k x (1 - \sin^2 x)^n \cos x \quad \text{and} \quad \cos^k x (1 - \cos^2 x)^n \sin x,$$

and expand; then each term is immediately integrable

because

$$\int \sin^m x \cos x\, dx = \frac{\sin^{m+1} x}{m+1} \quad (\S\ 38),$$

and

$$\int \cos^m x \sin x\, dx = -\frac{\cos^{m+1} x}{m+1} \, .$$

The same processes apply for $\sinh x$ and $\cosh x$.
By (v)

$$\int \tan x\, dx = \int \frac{\sin x}{\cos x}\, dx = -\log \cos x = \log \sec x \, ;$$

$$\int \cot x\, dx = \int \frac{\cos x}{\sin x}\, dx = \log \sin x \, ;$$

$$\int \tanh x\, dx = \log \cosh x \, ; \quad \int \coth x\, dx = \log \sinh x.$$

To integrate $\sec x$ and $\operatorname{cosec} x$, make use of the *substitution* $\tan \tfrac{1}{2}x = z$; then $x = 2\tan^{-1} z$, $dx = \dfrac{2dz}{1+z^2}$; also

$$\sin x = \frac{2z}{1+z^2}, \quad \cos x = \frac{1-z^2}{1+z^2}.$$

Then

$$\int \sec x\,dx = \int \frac{1+z^2}{1-z^2}\,\frac{2\,dz}{1+z^2}$$

$$= \int \frac{2\,dz}{1-z^2} = \int \left(\frac{1}{1+z} + \frac{1}{1-z}\right)dz$$

$$= \log \frac{1+z}{1-z}$$

$$= \log \frac{1+\tan \frac{1}{2}x}{1-\tan \frac{1}{2}x}$$

$$= \log \tan \left(\tfrac{1}{4}\pi + \tfrac{1}{2}x\right),$$

or $\quad = 2\tanh^{-1}\tan \frac{1}{2}x = \tanh^{-1}\sin x = -\cosh^{-1}\sec x.$

And, $\qquad \int \csc x\,dx = \int \frac{1+z^2}{2z}\,\frac{2\,dz}{1+z^2}$

$$= \int \frac{dz}{z} = \log z = \log \tan \tfrac{1}{2}x = -\cosh^{-1}\csc x.$$

Similarly, to integrate $\operatorname{sech} x$ and $\operatorname{cosech} x$, let $\tanh \frac{1}{2}x = z,$

then $dx = \dfrac{2\,dz}{1-z^2},$ and $\sinh x = \dfrac{2z}{1-z^2},$ $\cosh x = \dfrac{1+z^2}{1-z^2}.$

Then

$$\int \operatorname{sech} x\,dx = 2\tan^{-1}\tanh \tfrac{1}{2}x = \tan^{-1}\sinh x = \cos^{-1}\operatorname{sech} x.$$

$$\int \operatorname{cosech} x\,dx = \log \tanh \tfrac{1}{2}x = -\sinh^{-1}\operatorname{cosech} x.$$

The same substitution of z for $\tan \frac{1}{2}x$ or $\tanh \frac{1}{2}x$ will

reduce $\displaystyle \int \frac{dx}{a+b\cos x + c\sin x}$ or $\displaystyle \int \frac{dx}{a+b\cosh x + c\sinh x}$

to integrals of rational fractions; the different results will be found by reference to examples 85-89, page 57.

Examples.—Integrate with respect to x:

1. $\cos mx$, $\sin(mx+n)$, $\sin 2x \cos 3x$, $\cos 3x \cos 5x$, $\sin 3x \sin 5x$, $\sin(mx+n)\cos(px+q)$, $\sin x \sin 2x \sin 3x$, $\sin x \cos x$, $\sin^2 x \cos x$, $\sin^3 x \cos x$, $\sin^m x \cos x$, $\sin x \cos^2 x$, $\sin x \cos^3 x$, $\sin x \cos^m x$, $\sin^2 x$, $\sin^3 x$, $\sin^4 x$, $\cos^2 x$, $\cos^3 x$, $\cos^4 x$; also the same functions with $\sinh x$ for $\sin x$ and $\cosh x$ for $\cos x$.

2. $\tan x \sec^2 x$, $\tan^2 x \sec^2 x$, $\tan^m x \sec^2 x$, $\cot x \operatorname{cosec}^2 x$, $\cot^2 x \operatorname{cosec}^2 x$, $\cot^m x \operatorname{cosec}^2 x$, $\tan x$, $\cot x$, $\tan^2 x$, $\tan^3 x$, $\tan^4 x$, $\cot^2 x$, $\cot^3 x$, $\cot^4 x$, $\sec x \tan x$, $\sec^2 x \tan x$, $\sec^m x \tan x$, $\operatorname{cosec}^m x \cot x$, $\sec^2 x$, $\sec^4 x$, $\sec^6 x$, $\operatorname{cosec}^6 x$, $\sec x$, $\sec^3 x$, $\sec^5 x$, $\operatorname{cosec}^5 x$, $\sec x \operatorname{cosec} x$, $\operatorname{vers} x$, $\operatorname{vers}^2 x$; also the corresponding hyperbolic functions of x.

43. *Examples of integration by means of exponential, logarithmic, and inverse circular and hyperbolic functions*, using formulæ (i) to (q) and (v) to (z) of § 37.

Integrate with respect to x:

$$Ae^{ax+b}, \quad \frac{a}{bx}, \quad \frac{A}{x-a}, \quad \frac{2x+3}{x-1}, \quad \frac{(x-1)(x-2)}{x-3}, \quad \frac{Ax+B}{mx+n}, \quad \frac{2x+1}{x^2+x+1},$$

$$\frac{x^{n-1}}{a+bx^n}, \quad \frac{\cos x}{a+b\sin x}, \quad \frac{f'x}{a+bfx}, \quad \frac{1}{x}(\log x)^n, \quad \frac{1}{x(\log x)^n}, \quad \frac{1}{x \log x}, \quad \frac{1}{1+e^x},$$

$$\sqrt{(1+e^x)}, \quad \frac{1}{a\cos^2 x+b\sin^2 x}, \quad \frac{1}{a+b\tan x}, \quad \frac{1}{\sqrt{\{a^2-(x-c)^2\}}},$$

$$\frac{1}{\sqrt{\{(x-c)^2+a^2\}}}, \quad \frac{1}{\sqrt{\{(c-x)^2-a^2\}}}, \quad \frac{Px+Q}{\sqrt{(ax^2+2bx+c)}},$$

$$\frac{1}{a^2+(x-c)^2}, \quad \frac{1}{a^2-(x-c)^2}, \quad \frac{1}{x^2-x+1}, \quad \frac{2+x}{1+x+x^2}, \quad \frac{Px+Q}{ax^2+2bx+c},$$

$$\frac{2}{3+4x^2}, \quad \frac{x}{\sqrt{(x^2+a^2 . x^2+b^2)}}, \quad \frac{1}{(x^2+1)^{\frac{3}{2}}}, \quad \frac{1}{(x^4-1)^{\frac{1}{4}}}, \quad \frac{1}{(ax^n+b)^{1+n}}.$$

44. The Integral Calculus was invented for the purpose of finding areas, or for *quadrature*, as it was called, and the meaning of Integration is best illustrated by its application to finding the area of a curve.

Let $y = fx$ (fig. 9) be the equation of a curve CPQ; then if $OM = x$, $MP = fx$.

Let the area $AMPC$ between the curve and the axis of x bounded by an initial ordinate AC and the variable ordinate MP be denoted by A; keeping AC fixed and varying MP, A is some function of x, which it is required to determine.

Fig.9.

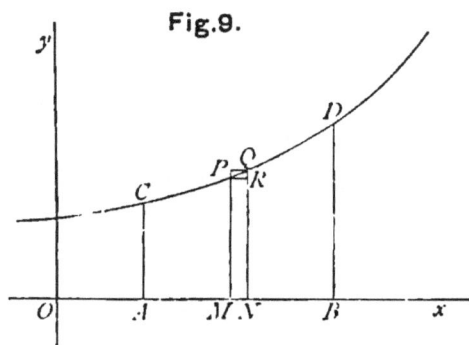

Let the ordinate MP be moved into the adjacent position NQ, by giving x the increment Δx; and let ΔA be the corresponding increment of A, and Δy of y.

Then $MN = \Delta x$, $NQ = y + \Delta y$; and the area $MNQP = \Delta A$.

But $MNQP$ lies between the rectangles PN and MQ, and therefore ΔA lies between $y\Delta x$ and $(y + \Delta y)\Delta x$; or, $\dfrac{\Delta A}{\Delta x}$ lies between y and $y + \Delta y$.

Proceeding to the limit by making Δx, and therefore Δy and ΔA indefinitely small,

$$\frac{d A}{d x} = \mathrm{lt}\frac{\Delta A}{\Delta x} = y = fx.$$

Therefore $\qquad$ A $=\int y dx +$ a constant

$$= \int f.x dx + \text{constant} ;$$

so that to determine A we must integrate f.x.

45. According to the Doctrine of Fluxions, the old-fashioned name of the Calculus, the area $AMPC$, called the *fluent*, is supposed to be generated by the flow of the variable ordinate MP; and the rate of increase of the area, called its *fluxion*, is equal to the ordinate MP into the fluxion of x; or, with our notation,

$$\frac{d\mathrm{A}}{dt} = y\frac{dx}{dt},$$

or

$$\frac{d\mathrm{A}}{dx} = y,$$

so that $\qquad$ A $=\int y dx.$

Similarly, any solid may be supposed generated by the motion of a variable plane area perpendicularly to itself; and then $\dfrac{dV}{dt} = A\dfrac{dx}{dt}$, if V denotes the volume, A the variable plane area, and $\dfrac{dx}{dt}$ the velocity of the plane perpendicular to itself.

For instance, if the volume V is bounded by the surface formed by the revolution of the curve $y = fx$ round the axis of x, then the fluent V may be supposed generated by the motion of an expanding (or contracting) circle of radius y, and therefore

$$\frac{dV}{dx} = \pi y^2, \text{ or } V = \pi\int y^2 dx.$$

46. Denoting the indefinite integral $\int fxdx$ by $f_{,}x$, then
$$A = f_{,}x - f_{,}a,$$
in order that A should vanish when $x = a$, supposing $OA = a$.

This is expressed in the notation of the Integral Calculus by

$$\int_a^x fxdx, \text{ or simply } \int_a fxdx = f_{,}x - f_{,}a,$$

a being called the *lower limit*, and the integral is then called a *corrected* integral.

47. Sometimes the fixed ordinate is taken to the right of the variable ordinate *MP*, at *BD* suppose, where $OB = b$; and then the area *MBDP* is expressed by

$$\int_x^b fxdx, \text{ or simply } \int^b fxdx = f_{,}b - f_{,}x,$$

and b is called the *upper* limit.

Thus, when corrected integrals, in § 37,

$$\int_a x^m dx = \frac{x^{m+1} - a^{m+1}}{m+1},$$

$$\int^b x^m dx = \frac{b^{m+1} - x^{m+1}}{m+1}.$$

$$\int \cos x dx = \sin x.$$

$$\int_0 \sin x dx = 1 - \cos x = \text{vers } x.$$

$$\int^{\frac{1}{2}\pi} \sin x dx = \cos x.$$

$$\int \sec^2 x dx = \tan x.$$

$$\int^{\frac{1}{2}\pi} \csc^2 x \, dx = \cot x.$$

$$\int_0 \frac{dx}{\sqrt{(a^2 - x^2)}} = \sin^{-1}\frac{x}{a}.$$

$$\int^a \frac{dx}{\sqrt{(a^2 - x^2)}} = \cos^{-1}\frac{x}{a}.$$

$$\int_0 \frac{dx}{a^2 + x^2} = \tan^{-1}\frac{x}{a}.$$

$$\int^x \frac{dx}{x^2 + a^2} = \cot^{-1}\frac{x}{a}.$$

$$\int_0 a^x dx = \frac{a^x - 1}{\log a}.$$

$$\int \frac{dx}{x} = \log \frac{x}{a}.$$

As an exercise, correct the remaining fundamental integrals of § **37** so that no constant shall be required.

48. Another geometrical interpretation of integration is here given, adapted from Newton's Lemma II., Principia, Lib. I., § 1.

Let the area $ABDC$ (fig. 10) bounded by the curve $y = fx$, the axis of x and the initial and final ordinates AC and BD, be divided into a large number of narrow strips like $PMNQ$ by equidistant ordinates at a distance Δx.

Then the difference between the external rectangle MQ and the internal rectangle PN is the rectangle PQ or pq; and therefore the difference between the sum of all the external rectangles and the sum of all the internal rectangles so described is the rectangle DL; also the area $ABDC$ is intermediate to the sum of the external and the sum of the internal rectangles.

Now in the limit when the breadth Δx of the rectangles is indefinitely diminished and their number proportionately increased, the rectangle DE vanishes,

Fig.10.

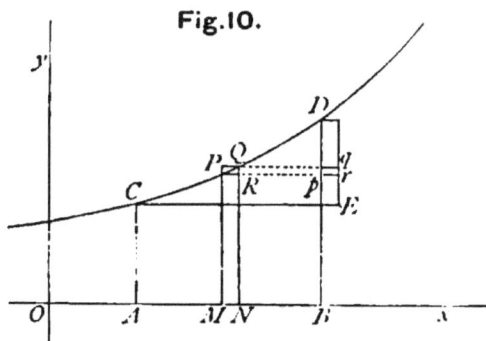

and therefore the sum of all the external and the sum of all the internal rectangles each become equal to the area $ABDC$.

If $OM = x$, $MP = y$, then the rectangle $NP = y\Delta x$; and denoting the sum of all the internal rectangles by $\Sigma y\Delta x$,

then the area

$$ABDC = \operatorname{lt}\sum_{x=a}^{x=b} y\Delta x,$$

$$= \int_{a}^{b} y\,dx = \int_{a}^{b} f x\,dx,$$

replacing Σ by $\int$, and Δx by dx in the limit, and supposing $OA = a$, $OB = b$.

49. Denoting the indefinite integral $\int f x\,dx$ by $f_1 x$, then

$$\int_{a}^{b} f x\,dx = f_1 b - f_1 a$$

and is called a *definite* integral; a being called the *lower* and b the *upper limit*.

The term *definite integral* is however retained *par excellence* for integrals which can only be evaluated between certain definite limits, and of which the indefinite integrals cannot be found.

Thus it can be proved that

$$\int_0^\infty e^{-x^2} dx = \sqrt{\pi},$$

$$\int_0^\infty \frac{\sin mx}{x} dx = \tfrac{1}{2}\pi \ ;$$

but the indefinite integrals

$$\int e^{-x^2} dx \ \text{ and } \ \int \frac{\sin mx}{x} dx$$

cannot be expressed by means of the algebraical, circular or hyperbolic functions we employ.

Examples.—Find the definite integrals:

$$\int_a^b x^n dx, \quad \int_0^{\frac{1}{2}\pi} \sin x dx, \quad \int_0^{\frac{1}{2}\pi} \cos x dx, \quad \int_0^{\frac{1}{4}\pi} \tan x dx, \quad \int_{\frac{1}{4}\pi}^{\frac{1}{2}\pi} \cot x dx,$$

$$\int_0^{\frac{1}{4}\pi} \sec x dx, \quad \int_{\frac{1}{4}\pi}^{\frac{1}{2}\pi} \operatorname{cosec} x dx, \quad \int_0^{\frac{1}{2}\pi} \sin^2 x dx, \quad \int_0^{\frac{1}{2}\pi} \sin^3 x dx,$$

$$\int_1^3 \frac{dx}{3+x^2}, \quad \int_0^\pi \frac{dx}{a + b \cos x} \quad (a > b).$$

50. We see that

$$\int_b^a f x dx = f_1 a - f_1 b = -\int_a^b f x dx,$$

so that an interchange of limits changes the sign of the definite integral.

Considerations of *symmetry* and *periodicity* of the function fx to be integrated are often useful.

Thus if fx is an *even* function of x, so that $f(-x) = fx$,

then
$$\int_{-a}^a f x dx = 2 \int_0^a f x dx \ ;$$

but if fx is an *odd* function, so that $f(-x) = -fx$, then

$$\int_{-a}^{a} fx\, dx = 0.$$

For instance, $\int_{\frac{1}{2}\pi}^{\frac{1}{2}\pi} \cos^{2n+1}x\, dx = 2\int_{0}^{\frac{1}{2}\pi} \cos^{2n+1}x\, dx,$

but $\int_{-\frac{1}{2}\pi}^{\frac{1}{2}\pi} \sin^{2n+1}x\, dx = 0;$

and $\int_{0}^{\pi} \sin^{2n+1}x\, dx = 2\int_{0}^{\frac{1}{2}\pi} \sin^{2n+1}x\, dx,$

but $\int_{0}^{\pi} \cos^{2n+1}x\, dx = 0.$

Again, if fx is a *periodic* function of *period l*, so that $f(x+nl) = fx$, where n is an integer, then

$$\int_{0}^{nl} fx\, dx = n\int_{0}^{l} fx\, dx.$$

It is advisable, in integration, to make use of these considerations in order to keep the limits of integration as close together as possible.

51. *Application of the Integral Calculus to the quadrature of the parabola* $y^2 = px$ (fig. 11).

The area OMP

$$= \int_{0} y\, dx = p^{\frac{1}{2}}\int_{0} x^{\frac{1}{2}} dx$$
$$= \tfrac{2}{3}p^{\frac{1}{2}}x^{\frac{3}{2}} = \tfrac{2}{3}xy$$
$$= \tfrac{2}{3} \text{ rectangle } OMPN.$$

Similarly, the area ONP

$$= \int_{0} x\, dy = \frac{1}{p}\int_{0} y^2\, dy$$
$$= \tfrac{1}{3}\frac{y^3}{p} = \tfrac{1}{3}xy$$
$$= \tfrac{1}{3} \text{ rectangle } OMPN.$$

52. The centre of gravity, or, as it is now called, the *centroid* G of the area OMP, where OP is any curve, can be determined by integration as follows : denoting the co-ordinates of the centroid G by $\bar{x}$, $\bar{y}$, then the moment of the whole area collected at G about the axes is equal

Fig. II.

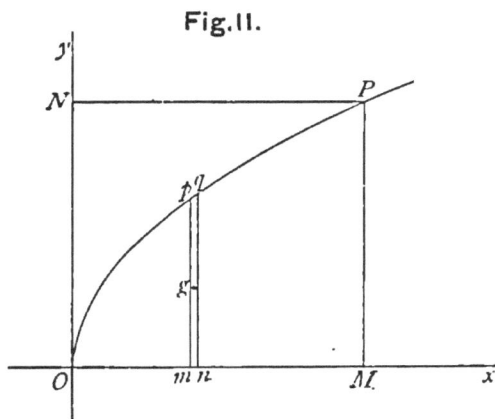

to the sum of the moments of the separate elements, such as mq, which may be supposed to have its centroid ultimately at g, the middle point of mp.

Therefore

$$\bar{x} = \frac{\int_0^{} xy\,dx}{\int_0^{} y\,dx}, \quad \bar{y} = \frac{\int_0^{} \tfrac{1}{2} y^2\,dx}{\int_0^{} y\,dx}.$$

It must be noticed in these integrals, as in all corrected integrals, that under the sign of integration x, y must be supposed to denote the co-ordinates Om, mp of any intermediate point p of the curve OP; but that when the integration is performed, then x, y represent OM, MP the co-ordinates of P, the variable upper limit of

F

the integration; this will not however be found to create confusion.

Generally, for the centroid G of the area $ABDC$ (fig. 10)

$$\bar{x} = \frac{\int_a^b xy\,dx}{\int_a^b y\,dx}, \quad \bar{y} = \frac{\int_a^b \tfrac{1}{2}y^2\,dx}{\int_a^b y\,dx}.$$

53. For the parabolic area OMP

$$\bar{x} = \frac{\int_0^{} x^{\frac{3}{2}}\,dx}{\int_0^{} x^{\frac{1}{2}}\,dx} = \frac{\tfrac{2}{5}x^{\frac{5}{2}}}{\tfrac{2}{3}x^{\frac{3}{2}}} = \tfrac{3}{5}x = \tfrac{3}{5}OM;$$

$$\bar{y} = \frac{\int_0^{} \tfrac{1}{2}px\,dx}{\int_0^{} p^{\frac{1}{2}}x^{\frac{1}{2}}\,dx} = \frac{\tfrac{1}{4}px^2}{\tfrac{2}{3}p^{\frac{1}{2}}x^{\frac{3}{2}}} = \tfrac{3}{8}p^{\frac{1}{2}}x^{\frac{1}{2}} = \tfrac{3}{8}MP.$$

For the centroid of the area ONP, taking y as the independent variable,

$$x = \frac{\int_0^{} \tfrac{1}{2}x^2\,dy}{\int_0^{} x\,dy} = \frac{\int_0^{} \frac{y^4}{2p^2}\,dy}{\int_0^{} \frac{y^2}{p}\,dy} = \frac{\frac{y^5}{10p^2}}{\frac{y^3}{3p}} = \frac{3}{10}\frac{y^2}{p} = \frac{3}{10}x;$$

$$\bar{y} = \frac{\int_0^{} xy\,dy}{\int_0^{} x\,dy} = \frac{\int_0^{} y^3\,dy}{\int_0^{} y^2\,dy} = \frac{\tfrac{1}{4}y^4}{\tfrac{1}{3}y^3} = \tfrac{3}{4}y.$$

Examples.—(1) Prove that if the equation of the curve OP (fig. 11) is $y^n = p^{n-m}x^m$, the area OMP is $\dfrac{nxy}{m+n}$, and the co-ordinates of its centroid are $\dfrac{m+n}{m+2n}x,\ \dfrac{m+n}{4m+2n}y$; and

that the area ONP is $\dfrac{mxy}{m+n}$, and the co-ordinates of its

centroid are $\dfrac{m+n}{2m+4n}x,\ \dfrac{m+n}{2m+n}y.$

(2) Prove that the area between the parabolas $y^2=px$ and $x^2=qy$ is $\frac{1}{3}pq$, and the co-ordinates of its centroid are $\bar{x}=\dfrac{9}{20}p^{\frac{1}{3}}q^{\frac{2}{3}},\ \bar{y}=\dfrac{9}{20}p^{\frac{2}{3}}q^{\frac{1}{3}}.$

(3) Prove that the area between $y^n=x^m$ and $y^m=x^n$ is $\dfrac{m-n}{m+n}$, and that the co-ordinates of its centroid are

$$\bar{x}=\bar{y}=\frac{(m+n)^2}{(m+2n)(2m+n)}.$$

54. *The quadrature of the circle and ellipse.*

In the circle (fig. 12),
$$x^2+y^2=a^2,$$
or $$y=\sqrt{(a^2-x^2)};$$
and therefore the area

$$OMPB=\int_0^x\sqrt{(a^2-x^2)}dx\ldots\ldots\ldots\ldots(\text{i.});$$

and the area

$$PMA=\int_x^a\sqrt{(a^2-x^2)}dx\ldots\ldots\ldots\ldots(\text{ii.}).$$

To find the first integral (i.),

put $\qquad x=a\sin\phi$, then the angle $BOP=\phi$;

and $\qquad y=\sqrt{(a^2-x^2)}=a\cos\phi,$

$\qquad\qquad dx=a\cos\phi d\phi.$

Therefore the area $OMPB$

$$= \int_0^\phi a^2\cos^2\phi\, d\phi$$

$$= \tfrac{1}{2}a^2\int_0^\phi (1+\cos 2\phi)\, d\phi$$

$$= \tfrac{1}{2}a^2(\phi + \tfrac{1}{2}\sin 2\phi)$$

$$= \tfrac{1}{2}a^2\phi + \tfrac{1}{2}a^2\sin\phi\cos\phi$$

of which the triangle $OMP = \tfrac{1}{2}a^2\sin\phi\cos\phi$, and the sector $OPB = \tfrac{1}{2}a^2\phi$.

Fig.12.

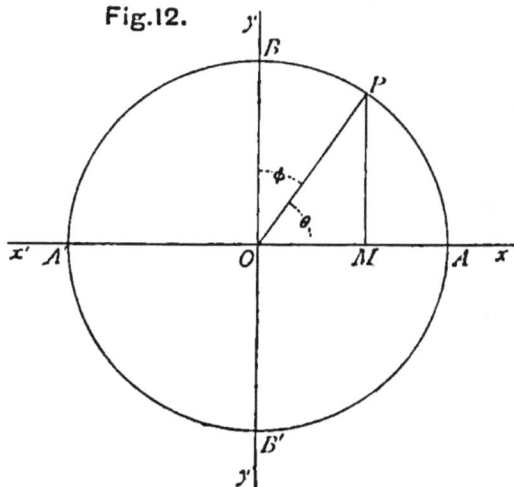

Expressed in terms of x, the area $OMPB$

$$= \int_0^x \sqrt{(a^2 - x^2)}\, dx$$

$$= \tfrac{1}{2}a^2\sin^{-1}\frac{x}{a} + \tfrac{1}{2}x\sqrt{(a^2 - x^2)}.$$

To find the second integral (ii.),

put $x = a\cos\theta$, then the angle $AOP = \theta$;

and $y = a\sin\theta$,

$$dx = -a\sin\theta\, d\theta.$$

Therefore the area PMA

$$= -\int_\theta^0 a^2 \sin^2\theta d\theta$$

$$= \tfrac{1}{2}a^2 \int_0^\theta (1 - \cos 2\theta) d\theta$$

$$= \tfrac{1}{2}a^2\theta - \tfrac{1}{2}a^2 \sin\theta \cos\theta;$$

the difference between the sector $OAP = \tfrac{1}{2}a^2\theta$, and the triangle $OMP = \tfrac{1}{2}a^2\sin\theta\cos\theta$.

Expressed in terms of x, the area PMA

$$= \int \sqrt{(a^2 - x^2)} dx$$

$$= \tfrac{1}{2}a^2 \cos^{-1}\frac{x}{a} - \tfrac{1}{2}x\sqrt{(a^2 - x^2)}.$$

Therefore the area of the quadrant OAB

$$= \int_0^a \sqrt{(a^2 - x^2)} dx = \tfrac{1}{4}\pi a^2;$$

and the area of the whole circle is πa^2.

For the centroid of the area $OMPB$

$$\bar{x} = \frac{\int x\sqrt{(a^2 - x^2)} dx}{\int_0^x \sqrt{(a^2 - x^2)} dx} = \frac{\tfrac{1}{3}a^3 - \tfrac{1}{3}(a^2 - x^2)^{\frac{3}{2}}}{\text{area } OMPB},$$

$$\bar{y} = \frac{\int_0^x \tfrac{1}{2}(a^2 - x^2) dx}{\int_0^x \sqrt{(a^2 - x^2)} dx} = \frac{\tfrac{1}{2}a^2 x - \tfrac{1}{6}x^3}{\text{area } OMPB},$$

For the centroid of the area PMA:

$$\bar{x} = \frac{\int x\sqrt{(a^2 - x^2)} dx}{\int_x^a \sqrt{(a^2 - x^2)} dx} = \frac{\tfrac{1}{3}(a^2 - x^2)^{\frac{3}{2}}}{\text{area } PMA},$$

$$\bar{y} = \frac{\int \frac{1}{2}(a^2 - x^2)dx}{\int^a \sqrt{(a^2 - x^2)}dx} = \frac{\frac{1}{2}a^2(a - x) - \frac{1}{6}(a^3 - x^3)}{\text{area } PMA}.$$

Therefore, for the quadrant OAB,

$$\bar{x} = \bar{y} = \frac{4a}{3\pi}.$$

If we take A as the origin and $AM = x$, $MP = y$; then $y = \sqrt{(2ax - x^2)}$; and the area $AMP = \int_0 \sqrt{(2ax - x^2)}dx$; which is reduced to the above by putting $x = a$ vers θ $= 2a \sin^2 \frac{1}{2}\theta$, and then $AOP = \theta$.

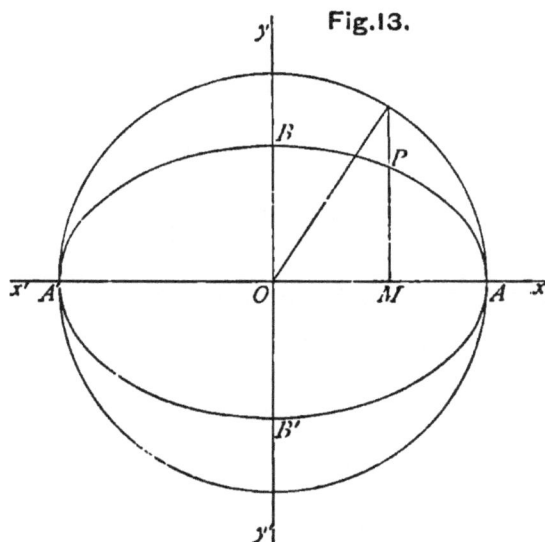

Fig.13.

55. The equation of the ellipse (fig. 13) is

$$\frac{x^2}{a^2} + \frac{y^2}{b^2} = 1,$$

or

$$y = \frac{b}{a}\sqrt{(a^2 - x^2)};$$

and therefore the area of a part of the ellipse is $\dfrac{b}{a}$ of the corresponding part of the auxiliary circle cut off by the same ordinate, θ being the excentric angle, and ϕ the complementary excentric angle.

Thus the sector OAP of the ellipse $= \frac{1}{2}ab\theta$, and the sector $OPB = \frac{1}{2}ab\phi$, the quadrant of the ellipse OAB being $= \frac{1}{4}\pi ab$; also the triangle

$$OMP = \tfrac{1}{2}ab \sin \theta \cos \theta = \tfrac{1}{2}ab \sin \phi \cos \phi.$$

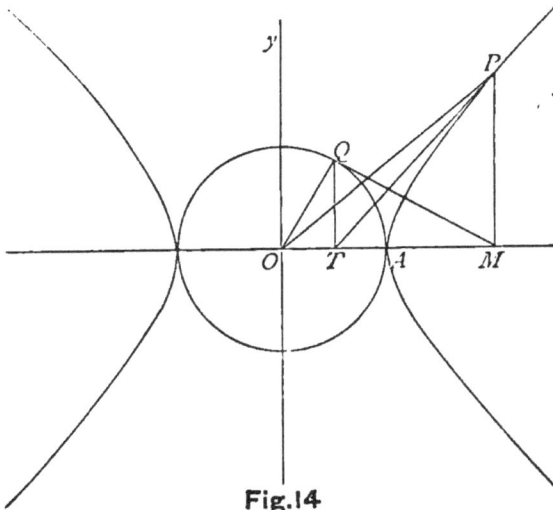

Fig.14

56. *The quadrature of the hyperbola and its conjugate.*

In the rectangular hyperbola (fig. 14)

$$x^2 - y^2 = a^2,$$

or
$$y = \sqrt{(x^2 - a^2)};$$

and therefore the area AMP

$$= \int_a \sqrt{(x^2 - a^2)}dx.$$

To find this integral by the hyperbolic functions, put

$$x = a \cosh u,$$

then

$$y = a \sinh u,$$

and

$$dx = a \sinh u\, du.$$

Therefore the area AMP

$$= \int_0 a^2 \sinh^2 u\, du$$
$$= \tfrac{1}{2}a^2 \int (\cosh 2u - 1) du$$
$$= \tfrac{1}{2}a^2(\tfrac{1}{2}\sinh 2u - u)$$
$$= \tfrac{1}{2}a^2 \sinh u \cosh u - \tfrac{1}{2}a^2 u,$$

of which the triangle $OMP = \tfrac{1}{2}a^2 \sinh u \cosh u$, and therefore the sector $OAP = \tfrac{1}{2}a^2 u$ (§ 30).

Expressed in terms of x, the area AMP

$$= \int_a \sqrt{(x^2 - a^2)} dx$$
$$= \tfrac{1}{2}x \sqrt{(x^2 - a^2)} - \tfrac{1}{2}a^2 \cosh^{-1}\frac{x}{a}$$
$$= \tfrac{1}{2}x \sqrt{(x^2 - a^2)} - \tfrac{1}{2}a^2 \log \frac{x + \sqrt{(x^2 - a^2)}}{a}.$$

To find the integral by the circular functions, put $x = a \sec \theta$, then $y = a \tan \theta$, and the angle $AOQ = \theta$.

Then the area AMP

$$= a^2 \int_0 \tan^2 \theta \sec \theta\, d\theta,$$
$$= \tfrac{1}{2}a^2 \sec \theta \tan \theta - \tfrac{1}{2}a^2 \log (\sec \theta + \tan \theta)$$
$$= \tfrac{1}{2}x \sqrt{(x^2 - a^2)} - \tfrac{1}{2}a^2 \log \frac{x + \sqrt{(x^2 - a^2)}}{a}$$

as before.

If we take A as the origin, and put $AM = x$, then
$$MP = y = \sqrt{(2ax + x^2)},$$
and the area
$$AMP = \int_0 \sqrt{(2ax + x^2)}\,dx,$$

which is reduced by putting
$$x = 2a\sinh^2\tfrac{1}{2}u, \quad 2a + x = 2a\cosh^2\tfrac{1}{2}u;$$
then
$$y = a\sinh u,$$
and
$$dx = 2a\sinh\tfrac{1}{2}u\cosh\tfrac{1}{2}u\,du$$
$$= a\sinh u\,du;$$
so that the area
$$AMP = a^2\int_0 \sinh^2 u\,du$$

as before.

57. In the conjugate rectangular hyperbola (fig. 15),
$$y^2 - x^2 = a^2,$$
or
$$y = \sqrt{(a^2 + x^2)}.$$

Then the area $OMPB$
$$= \int_0 \sqrt{(a^2 + x^2)}\,dx,$$

which is reduced to $a^2\int_0 \cosh^2 v\,dv$ by putting $x = a\sinh v$, $y = a\cosh v$, and then the sector $OBP = \tfrac{1}{2}a^2 v$; or the integral is reduced to $a^2\int_0 \sec^3\phi\,d\phi$ by putting $x = a\tan\phi$, $y = a\sec\phi$, and then the angle $BOQ = \phi$.

Therefore the area $OMPB$
$$= \int_0 \sqrt{(a^2 + x^2)}\,dx$$
$$= \tfrac{1}{2}x\sqrt{(a^2 + x^2)} + \tfrac{1}{2}a^2\sinh^{-1}\frac{x}{a}$$
$$= \tfrac{1}{2}x\sqrt{(a^2 + x^2)} + \tfrac{1}{2}a^2\log\frac{\sqrt{(a^2 + x^2)} + x}{a};$$

and its centroid is given by

$$\bar{x}=\frac{\int_0^{x}\sqrt{(a^2+x^2)}dx}{\int_0^{}\sqrt{(a^2+x^2)}dx}=\frac{\frac{1}{3}(a^2+x^2)^{\frac{3}{2}}-\frac{1}{3}a^3}{\text{area } OMPB},$$

$$\bar{y}=\frac{\int_0^{}\frac{1}{2}(a^2+x^2)dx}{\int_0^{}\sqrt{(a^2+x^2)}dx}=\frac{\frac{1}{2}a^2x+\frac{1}{6}x^3}{\text{area } OMPB}.$$

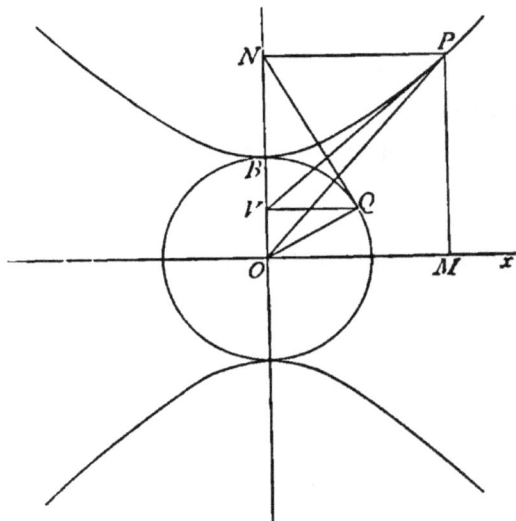

Fig.15

58. The corresponding properties of the hyperbola

$$\frac{x^2}{a^2}-\frac{y^2}{b^2}=1,$$

and of its conjugate hyperbola,

$$\frac{y^2}{b^2}-\frac{x^2}{a^2}=1,$$

are obtained immediately by projecting the preceding figures orthogonally on a plane parallel to Ox; and then the co-ordinates of any point P on the hyperbola are

$$x=a\cosh u, \quad y=b\sinh u,$$

where the sectorial area OAP (fig. 14)

$$= \tfrac{1}{2}abu = \tfrac{1}{2}ab\log\left(\frac{x}{a}+\frac{y}{b}\right);$$

or we may put $x = a\sec\theta,\ y = b\tan\theta.$

The co-ordinates of any point P on the conjugate hyperbola are $x = a\sinh v,\ y = b\cosh v,$ where the sectorial area OBP (fig. 15)

$$= \tfrac{1}{2}abv = \tfrac{1}{2}ab\log\left(\frac{x}{a}+\frac{y}{b}\right);$$

or we may put $x = a\tan\phi,\ y = b\sec\phi.$

59. Any integral of the form $\displaystyle\int\frac{x^m dx}{\sqrt{(px^2+2qx+r)}}$ can be

reduced by means of one of the above substitutions in connection with the ellipse, the hyperbola or its conjugate, according to the form of the conic section

$$y = \sqrt{(px^2+2qx+r)}.$$

Examples.—Draw the following curves, and denoting the upper half of the area to the right of the axis of y by A and the co-ordinates of its centroid by $\bar{x},\ \bar{y}$, prove that

(i.) In $ay^2 = ax^2 - x^3,\ A = \dfrac{4}{15}a^2,\ \bar{x} = \dfrac{4}{7}a,\ \bar{y} = \dfrac{5}{32}a$;

(ii.) In $a^2y^2 = ax^3 - x^4,\ A = \dfrac{\pi a^2}{16},\ \bar{x} = \dfrac{5}{8}a,\ \bar{y} = \dfrac{2a}{5\pi}$;

(iii.) In $xy^2 = a^3 - a^2x,\ A = \tfrac{1}{2}\pi a^2,\ \bar{x} = \tfrac{1}{4}a,\ \bar{y} = \infty$;

(iv.) In $a^2y^2 = x^2(a^2 - x^2),\ A = \tfrac{1}{3}a^2,\ \bar{x} = \tfrac{3}{16}\pi a,\ \bar{y} = \tfrac{1}{5}a.$

60. *Quadrature with Polar Co-ordinates.*

Let $r = f\theta$ be the polar equation of a curve BPQ (fig. 16), then if $xOP = \theta$, $OP = r = f\theta$.

Let the sectorial area OBP enclosed by the curve BP, an initial vector OB, and the variable vector OP be denoted by A.

If ΔA and Δr denote the increments of A and r corre-

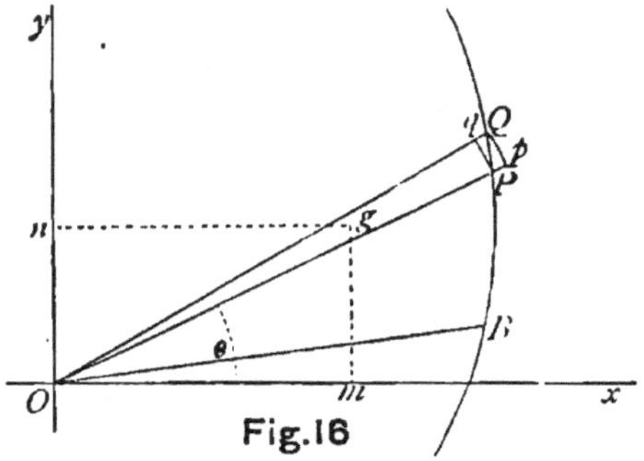

Fig.16

sponding to the increment $\Delta\theta$ of θ, then if $POQ = \Delta\theta$, $xOQ = \theta + \Delta\theta$, $OQ = r + \Delta r = f(\theta + \Delta\theta)$, and the area $OPQ = \Delta A$.

Drawing the circular arcs Pq, Qp, with centre O, the sectorial area OPQ is seen to be intermediate to the circular sectors OPq and OpQ; or ΔA lies between $\frac{1}{2}r^2\Delta\theta$ and $\frac{1}{2}(r + \Delta r)^2\Delta\theta$, since the circular sector $OPq = \frac{1}{2}r^2\Delta\theta$, and the circular sector $OpQ = \frac{1}{2}(r + \Delta r)^2\Delta\theta$; and therefore $\dfrac{\Delta A}{\Delta\theta}$ lies between $\frac{1}{2}r^2$ and $\frac{1}{2}(r + \Delta r)^2$.

Proceeding to the limit, by making $\Delta\theta$ indefinitely small,

$$\frac{dA}{d\theta} = \tfrac{1}{2}r^2.$$

Therefore $A = \frac{1}{2}\int_a^\theta r^2 d\theta = \frac{1}{2}\int_a (f\theta)^2 d\theta,$

if the angle $xOB = \alpha$, the formula to be employed in finding areas with polar co-ordinates.

Also if $\bar{x}$, $\bar{y}$ denote the co-ordinates of the centroid of the area OBP, then since g the centroid of the element OPQ is ultimately in OP at a distance $\frac{2}{3}r$ from O, because OPQ may be considered ultimately a triangle, therefore

$$\bar{x} = \frac{\int_\alpha \frac{2}{3}r\cos\theta\frac{1}{2}r^2 d\theta}{\int_\alpha \frac{1}{2}r^2 d\theta} = \frac{\frac{1}{3}\int_\alpha r^3\cos\theta d\theta}{A},$$

$$\bar{y} = \frac{\int_\alpha \frac{2}{3}r\sin\theta\frac{1}{2}r^2 d\theta}{\int_\alpha \frac{1}{2}r^2 d\theta} = \frac{\frac{1}{3}\int_\alpha r^3\sin\theta d\theta}{A},$$

where $r = f\theta$.

Examples.—(1) Find the area of a circle when its equation is given in the form

$$r = 2a\cos\theta.$$

Here
$$A = \frac{1}{2}\int_{-\frac{1}{2}\pi}^{\frac{1}{2}\pi} r^2 d\theta = 2a^2\int\cos^2\theta d\theta$$

$$= a^2\int_{-\frac{1}{2}\pi}^{\frac{1}{2}\pi}(1+\cos 2\theta)d\theta = \pi a^2.$$

(2) Find the area and centroid of a loop of the curve $r = a\cos 2\theta$.

Answer: $A = \frac{1}{8}\pi a^2$, $\bar{x} = \frac{128\sqrt{2}}{105\pi}a.$

(3) Find the area of a loop of the curve

$$r = a\cos n\theta + b\sin n\theta.$$

Answer: $A = \frac{\pi}{4n}(a^2 + b^2).$

(4) Find the area and centroid of a loop of the curve $r^2 = a^2\cos 2\theta$.

Answer: $\qquad A = \frac{1}{2}a^2, \; \bar{x} = \frac{1}{8}\pi a\sqrt{2}.$

(5) Find the area and centroid of $r = a(1 + \cos\theta)$.

Answer: $\qquad A = \frac{3}{2}\pi a^2, \; \bar{x} = \frac{5}{6}a.$

(6) Find the centroid of a sector OAP of a circle (fig. 12) of radius a.

Here
$$\bar{x} = \frac{\int_0^{\theta} \frac{2}{3}a\cos\theta \frac{1}{2}a^2 d\theta}{\int_0^{\theta} \frac{1}{2}a^2 d\theta} = \frac{2}{3}a\frac{\sin\theta}{\theta},$$

$$\bar{y} = \frac{\int_0^{\theta} \frac{2}{3}a\sin\theta \frac{1}{2}a^2 d\theta}{\int_0^{\theta} \frac{1}{2}a^2 d\theta} = \frac{2}{3}a\frac{1 - \cos\theta}{\theta}.$$

Similarly for the centroid of a sector OAP of an ellipse (fig. 13).

$$\bar{x} = \frac{2}{3}a\frac{\sin\theta}{\theta}, \; \bar{y} = \frac{2}{3}b\frac{1 - \cos\theta}{\theta},$$

where θ is the excentric angle of the point P on the ellipse.

(7) Find the centroid of a sector OAP of a hyperbola (fig. 14).

Here
$$A = \frac{1}{2}abu, \; dA = \frac{1}{2}abdu,$$

$$\bar{x} = \frac{\int_0^{\theta} \frac{2}{3}a\cosh u\frac{1}{2}abdu}{\int_0^{\theta} \frac{1}{2}abdu} = \frac{2}{3}a\frac{\sinh u}{u},$$

$$\bar{y} = \frac{\int_0^{\theta} \frac{2}{3}a\sinh u\frac{1}{2}abdu}{\int_0^{\theta} \frac{1}{2}abdu} = \frac{2}{3}b\frac{\cosh u - 1}{u}.$$

61. *Rectification of Curves.*

The formulæ of the Differential Calculus

$$\frac{ds^2}{dt^2} = \frac{dx^2}{dt^2} + \frac{dy^2}{dt^2} \quad (\S\ 13),$$

$$\frac{ds^2}{dt^2} = \frac{dr^2}{dt^2} + \frac{r^2 d\theta^2}{dt^2} \quad (\S\ 18),$$

become in the Integral Calculus

$$s = \int \sqrt{\left(\frac{dx^2}{dt^2} + \frac{dy^2}{dt^2}\right)} dt,$$

$$s = \int \sqrt{\left(\frac{dr^2}{dt^2} + \frac{r^2 d\theta^2}{dt^2}\right)} dt\ ;$$

or

$$s = \int \sqrt{(dx^2 + dy^2)},$$

$$s = \int \sqrt{(dr^2 + r^2 d\theta^2)},$$

leaving the independent variable arbitrary.

Also if $\bar{x}$, $\bar{y}$ are the co-ordinates of the centroid or centre of mass of a material curve or wire, of density σ per unit of length,

$$\bar{x} = \frac{\int \sigma x\, ds}{\int \sigma\, ds},\ \bar{y} = \frac{\int \sigma y\, ds}{\int \sigma\, ds}.$$

Examples.—Rectify the following curves—that is, find s in terms of x or y, or in polar co-ordinates in terms of θ or r. It is supposed that s is measured from the point where $x = 0$, or $\theta = 0$.

(1) The *semi-cubical* parabola $9ay^2 = 4x^3$.

Here
$$\frac{ds}{dx} = \sqrt{\left(1 + \frac{x}{a}\right)},$$

$$s = \int_0 \sqrt{\left(1 + \frac{x}{a}\right)} dx$$

$$= \tfrac{2}{3}a \left\{ \left(1 + \frac{x}{a}\right)^{\frac{3}{2}} - 1 \right\}.$$

(2) To find when a curve whose equation is of the form $y = x^r$ is rectifiable.

Here $$s = \int \sqrt{(1 + r^2 x^{2r-2})} dx,$$

which is integrable when r is of the form $\dfrac{2i+1}{2i}$, or $\dfrac{2i}{2i-1}$, where i is any integer.

For, consider the more general integral,

$$\int x^{m-1}(a + bx^n)^{\frac{p}{q}} dx;$$

this integral is rationalized when $\dfrac{m}{n}$ is an integer by means of the substitution $a + bx^n = z^q$, and when $\dfrac{m}{n} + \dfrac{p}{q}$ is an integer by the substitution $a + bx^n = x^n z^q$.

(3) In $x^{\frac{2}{3}} + y^{\frac{2}{3}} = a^{\frac{2}{3}}$, prove that $s = \frac{3}{2} a^{\frac{1}{3}} x^{\frac{2}{3}}$.

(4) In the catenary $y = a \cosh \dfrac{x}{a}$, $s = a \sinh \dfrac{x}{a}$.

(5) In the tractrix (ex. 92, p. 58) prove that $s = a \log \dfrac{a}{y}$.

(6) Prove that in the curve $y = a \log \sec \dfrac{x}{a}$ (ex. 90, p. 57)

$$\cosh \frac{s}{a} = \sec \frac{x}{a}, \text{ or } \frac{x}{a} = \gd \frac{s}{a}.$$

(7) With polar co-ordinates, prove that in the curve $r = a \cos \theta$, $s = a\theta$.

(8) In $r = a(1 + \cos \theta)$ (the *cardioid*), $s = 4a \sin \frac{1}{2}\theta$.

(9) In $r = \dfrac{2a}{1 + \cos \theta}$ (the *parabola*),

$$s = a \sec \tfrac{1}{2}\theta \tan \tfrac{1}{2}\theta + a \log (\sec \tfrac{1}{2}\theta + \tan \tfrac{1}{2}\theta).$$

62. In the parabola $y^2 = 2lx$, taking y as the independent variable,

$$s = \int_0 \sqrt{\left(\frac{dx^2}{dy^2} + 1\right)} dy$$

$$= \int_0 \sqrt{\left(\frac{y^2}{l^2} + 1\right)} dy$$

$$= \tfrac{1}{2} y \sqrt{\left(\frac{y^2}{l^2} + 1\right)} + \tfrac{1}{2} l \sinh^{-1} \frac{y}{l}.$$

Again, in the curve $r = l\theta$ (the *spiral of Archimedes*), taking r as the independent variable

$$s = \int_0 \sqrt{\left(1 + \frac{r^2 d\theta^2}{dr^2}\right)} dr$$

$$= \int_0 \sqrt{\left(1 + \frac{r^2}{l^2}\right)} dr$$

$$= \tfrac{1}{2} r \sqrt{\left(1 + \frac{r^2}{l^2}\right)} + \tfrac{1}{2} l \sinh^{-1} \frac{r}{l},$$

the same function of r as in the parabola s is of y.

Consequently a cylinder whose cross section is the

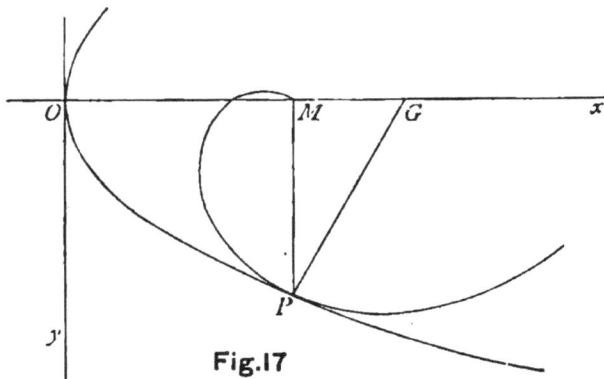

Fig.17

curve $r = l\theta$, loaded so that the centre of gravity is at the pole M, can be placed on the parabola $y^2 = 2lx$ with axis horizontal as in fig. 17, so as always to rest in neutral equilibrium ; and MG, the subnormal $y\dfrac{dy}{dx}$ of the parabola, and the polar subnormal $\dfrac{dr}{d\theta}$ of the spiral are each l.

Examples.—(1) Work out the corresponding problem for the ellipse $\dfrac{x^2}{a^2}+\dfrac{y^2}{b^2}=1$, and the curve $r=b\cos\dfrac{b}{a}\theta$.

(2) Also for the *exponential* or *logarithmic* curve $y=be^{\frac{x}{a}}$ and the *hyperbolic* spiral $r\theta=c$.

(3) Also for the *trochoid*, a curve traced by a point at a distance b from the centre of a circle of radius a which rolls on a straight line, and the *limaçon* $r=a+b\cos\theta$.

(4) Also for the catenary $y=a\cosh\dfrac{x}{a}$ and a straight line.

(5) Also for a straight line inclined at an angle a to the horizon, and the *equiangular* or *logarithmic* *spiral* $r=ae^{\theta\tan a}$.

(6) Also for the curve $y=a\left(1+e\cos\dfrac{x}{b}\right)$, and an ellipse with centre of gravity at a focus.

63. *To find the volume and surface of a solid of revolution.*

It has already been shown (§ 45) that if the volume V is contained by the surface made by the revolution of the curve $y=fx$ round the axis of x and by planes perpendicular to the axis, then

$$\frac{dV}{dx}=\pi y^2=\pi(fx)^2,$$

and $V=\pi\int y^2dx=\pi\int(fx)^2dx.$

Also if $\bar{x}$ is the abscissa of the centroid of the solid V, the centroid being in the axis of revolution,

$$\bar{x} = \frac{\pi\int xy^2dx}{\pi\int y^2dx} = \frac{\int x(fx)^2dx}{\int (fx)^2dx}.$$

Again, if S denotes the surface generated by the revolution of the curve $y = fx$, then the *fluent* S may be supposed generated by the motion of an expanding circumference of radius y, and therefore

$$\frac{dS}{dt} = 2\pi y\frac{ds}{dt}, \text{ or } \frac{dS}{dx} = 2\pi y\frac{ds}{dx}.$$

Therefore $\quad S = 2\pi\int y\frac{ds}{dx}dx = 2\pi\int yds,$

and for the centroid of the surface

$$\bar{x} = \frac{2\pi\int xy\frac{ds}{dx}dx}{2\pi\int y\frac{ds}{dx}dx} = \frac{\int xyds}{\int yds}.$$

64. *Application to the Sphere, Spheroid, Paraboloid, and Cone.*

(i.) In the *sphere*, generated by the revolution of a circle (fig. 18) round the axis of x,

$$y^2 = a^2 - x^2,$$

and therefore

$$V = \pi\int_0 (a^2 - x^2)dx = \pi(a^2x - \tfrac{1}{3}x^3),$$

$$\bar{x} = \frac{\pi\int_0 (a^2x - x^3)dx}{\pi\int (a^2 - x^2)dx} = \frac{\tfrac{1}{2}a^2x^2 - \tfrac{1}{4}x^4}{a^2x - \tfrac{1}{3}x^3}.$$

For a hemisphere, therefore,

$$V = \tfrac{2}{3}\pi a^3, \text{ and } \bar{x} = \tfrac{3}{8}a.$$

Again, since

$$\frac{ds}{dx} = \frac{a}{y},$$

$$S = 2\pi \int_0^a y\frac{a}{y}dx = 2\pi a x,$$

$$\bar{x} = \frac{\int_0^a x\,dx}{\int_0^a dx} = \frac{\tfrac{1}{2}x^2}{x} = \tfrac{1}{2}x.$$

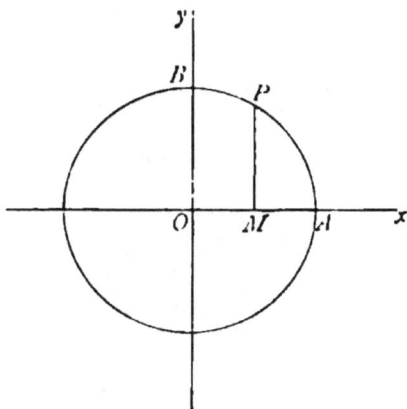

Fig. 18.

(ii.) In the *spheroid*, generated by the revolution of an ellipse round the axis of x,

$$y^2 = \frac{b^2}{a^2}(a^2 - x^2)\ ;$$

and therefore

$$V = \pi\frac{b^2}{a^2}(a^2 x - \tfrac{1}{3}x^3) = \pi a b^2\left(\frac{x}{a} - \tfrac{1}{3}\frac{x^3}{a^3}\right)\ ;$$

and for the hemispheroid,

$$V = \tfrac{2}{3}\pi a b^2,\ \bar{x} = \tfrac{3}{8}a.$$

If the spheroid is prolate, like a lemon, a is greater than b; if oblate, like an orange, a is less than b.

(iii.) In the *paraboloid*, generated by the revolution of the parabola $y^2 = 2lx$,

$$V = \pi \int_0^{} 2lx\,dx = \pi l x^2 = \tfrac{1}{2}\pi x y^2$$

$= \tfrac{1}{2}$ volume of the circumscribing cylinder;

and $\qquad \bar{x} = \tfrac{2}{3}x.$

Also $\qquad S = 2\pi \int_0^{} y \dfrac{ds}{dy} dy$

$$= 2\pi \int_0^{} y \sqrt{\left(\dfrac{y^2}{l^2} + 1 \right)} dy$$

$$= \tfrac{2}{3}\pi l^2 \left\{ \left(\dfrac{y^2}{l^2} + 1 \right)^{\frac{3}{2}} - 1 \right\}.$$

(iv.) In the *cone*, generated by the revolution of the straight line $y = x \tan a$, where a is the semi-vertical angle of the cone,

$$V = \pi \int_0^{} x^2 \tan^2 a\,dx = \tfrac{1}{3}\pi x^3 \tan^2 a = \tfrac{1}{3}\pi x y^2$$

$= \tfrac{1}{3}$ volume of the circumscribing cylinder;

and $\qquad \bar{x} = \tfrac{3}{4}x.$

Also $\qquad S = 2\pi \int_0^{} y \operatorname{cosec} a\,dy = \pi y^2 \operatorname{cosec} a;$

and for the centroid of the surface S,

$$\bar{x} = \tfrac{2}{3}x.$$

65. *Theorems of Pappus (or Guldin).*

Theorem I.—The volume V generated by the revolution of a closed plane curve of area A about an axis in its plane is equal to the volume of a cylinder of base A, and height equal to the circumference of the circle described by the centroid of the area A.

First consider the volume generated by the revolution of the area $ABDC$ in fig. 10; then

$$V = \pi \int y^2 dx;$$

and

$$\bar{y}A = \tfrac{1}{2} \int y^2 dx,$$

if A denotes the area of $ABDC$ and $\bar{y}$ the ordinate of G, the centroid of the area.

Therefore $V = 2\pi\bar{y}A$, which proves the theorem.

More generally, if A denotes the area of any closed plane curve not intersected by the axis of x, and if $\bar{y}$ denotes the ordinate of its centroid, then supposing the area A to be cut up into elementary strips by straight lines parallel to the axis of x, the length of a strip at a distance y from the axis of x will be some function $\phi(y)$ of y; and

$$A = \int \phi(y)dy, \quad \bar{y}A = \int y\phi(y)dy;$$

and

$$V = 2\pi \int y\phi(y)dy;$$

so that

$$V = 2\pi\bar{y}A, \text{ as before.}$$

Theorem II.—The surface S generated by the revolution of a plane curve of length s about an axis in its plane is equal to a rectangle of base s and height equal to the circumference of the circle described by the centroid of the arc s.

For if $\bar{y}$ denotes the ordinate of the centroid of the arc CD (fig. 10), $ys = \int y\,ds$;

also $$S = 2\pi \int y\,ds,$$

so that $$S = 2\pi \bar{y}s,$$

which proves the theorem.

Examples.—(1) Prove that in the figure called the *anchor ring*, made by the revolution of a circle (of radius c) about an axis in the plane of the circle (at a distance a from the centre),

$$V = 2\pi^2 ac^2, \ S = 4\pi^2 ac.$$

(2) Prove that the volume of a *parabolic spindle* made by the revolution of a parabola about an ordinate is $\frac{8}{15}$ of the volume of the circumscribing cylinder.

(3) Determine by Pappus's Theorems the centroid of the area and of the arc of a semicircle, knowing the volume and surface of a sphere.

(4) Determine the volumes and surfaces generated by the revolution about the co-ordinate axes of the curves in the examples of §§ 59, 60, and 61.

66. *Integration by Parts.*

Corresponding to the formula in the Differential Calculus for the differentiation of a product (§ 34),

$$\frac{duv}{dx} = \frac{du}{dx}v + u\frac{dv}{dx},$$

we have, integrating both sides,

$$uv = \int \frac{du}{dx}v\,dx + \int u\frac{dv}{dx}\,dx,$$

or $$\int u\frac{dv}{dx}\,dx = uv - \int \frac{du}{dx}v\,dx;$$

or, as it may be written,

$$\int u\,dv = uv - \int v\,du,$$

the formula of the Integral Calculus, for *integration by parts*, as it is called.

The formula shows how $\int u\,dv$ is made to depend upon $\int v\,du$, which is either integrable, or made to depend upon another integral, by another application of the method of integration by parts.

Thus to integrate

(i.) xe^x, suppose $x = u$, $\dfrac{dv}{dx} = e^x$, then $v = e^x$ and

$$\int xe^x dx = xe^x - \int e^x dx$$
$$= xe^x - e^x.$$

(ii.) $\log x$, suppose $\log x = u$, $\dfrac{dv}{dx} = 1$, then $v = x$,

and $\quad\displaystyle\int \log x\,dx = x \log x - \int x \frac{dx}{x} = x \log x - x.$

Similarly $\quad\displaystyle\int x^m \log x\,dx = \frac{x^{m+1}\log x}{m+1} - \frac{x^{m+1}}{(m+1)^2}.$

(iii.) $\sin\sqrt{x}$, first put $\sqrt{x} = z$, then

$$\int \sin\sqrt{x}\,dx = 2\int z \sin z\,dz$$
$$= -2z \cos z + 2\int \cos z\,dz$$
$$= -2z \cos z + 2 \sin z$$
$$= -2\sqrt{x} \cos \sqrt{x} + 2 \sin \sqrt{x}.$$

(iv.) To integrate $\cos mx \cosh nx$; here $\cos mx$ or $\cosh nx$ may be taken indiscriminately for u, and we must integrate by parts twice.

For instance, if $\cos mx = u$, $v = \dfrac{1}{n}\sinh nx$,

$$\int \cos mx \cosh nx\, dx = \frac{1}{n}\cos mx \sinh nx + \frac{m}{n}\int \sin mx \sinh nx\, dx$$

$$= \frac{1}{n}\cos mx \sinh nx + \frac{m}{n^2}\sin mx \cosh nx - \frac{m^2}{n^2}\int \cos mx \cosh nx\, dx.$$

Therefore

$$\int \cos mx \cosh nx\, dx = \frac{m \sin mx \cosh nx + n \cos mx \sinh nx}{m^2 + n^2}.$$

Similarly

$$\int \sin mx \sinh nx\, dx = \frac{n \sin mx \cosh nx - m \cos mx \sinh nx}{m^2 + n^2},$$

$$\int \cos mx \sinh nx\, dx = \frac{n \cos mx \cosh nx + m \sin mx \sinh nx}{m^2 + n^2},$$

$$\int \sin mx \cosh nx\, dx = \frac{-m \cos mx \cosh nx + n \sin mx \sinh nx}{m^2 + n^2};$$

and $\quad\displaystyle\int e^{nx}\cos mx\, dx = e^{nx}\frac{m \sin mx + n \cos mx}{m^2 + n^2},$

$$\int e^{nx}\sin mx\, dx = e^{nx}\frac{n \sin mx - m \cos mx}{m^2 + n^2}.$$

Examples.—Integrate by parts:

$x^2 e^x$, $x^3 e^{ax}$, $x^4 e^{ax+b}$, $x^m \log x^n$, $(\log x)^2$, $(\log x)^3$, $\cos x^{\frac{1}{3}}$, $\sin x^{\frac{1}{4}}$, $x^2\cos x$, $x^3\sin x$, $e^{ax+b}\cos(mx+n)$, $e^{ax+b}\sin(mx+n)$, $\sqrt{(a^2 - x^2)}$, $\sqrt{(x^2 - a^2)}$, $\sqrt{(a^2 + x^2)}$, $(a^2 - x^2)^{\frac{3}{2}}$, $x^2\sqrt{(a^2 + x^2)}$, $\sin^{-1}x$, $\cos^{-1}x$, $\tan^{-1}x$, $\cot^{-1}x$, $\sec^{-1}x$, $\text{vers}^{-1}x$, $\sinh^{-1}x$, $\cosh^{-1}x$, $\tanh^{-1}x$, $\text{sech}^{-1}x$, $(\sin^{-1}x)^2$, $(\text{vers}^{-1}x)^2$, $(\sinh^{-1}x)^2$, $(\cosh^{-1}x)^2$.

67. *Formulæ of Reduction.*

A *formula of reduction* in the Integral Calculus is a formula obtained in general by integration by parts, by means of which one integral, say u_m, is made to depend upon a simpler integral u_{n-1} or u_{n-2}; then by successive substitution in the formula of reduction we finally arrive at an integration which can be effected.

For instance, suppose $u_n = \int (\sin x)^n dx$; integrating by parts,

$$\int (\sin x)^n dx = \int (\sin x)^{n-1} \sin x dx$$

$$= -(\sin x)^{n-1}\cos x + (n-1)\int (\sin x)^{n-2} \cos^2 x dx$$

$$= -(\sin x)^{n-1}\cos x + (n-1)\int \{(\sin x)^{n-2} - (\sin x)^n\} dx$$

or $\quad n\int (\sin x)^n dx = -(\sin x)^{n-2}\cos x + (n-1)\int (\sin x)^{n-1} dx,$

or $\quad u_n = -\dfrac{1}{n}(\sin x)^{n-1}\cos x + \dfrac{n-1}{n} u_{n-2},$

a formula of reduction.

Similarly

$$\int (\cos x)^n dx = \frac{1}{n}\sin x (\cos x)^{n-1} + \frac{n-1}{n}\int (\cos x)^{n-2} dx,$$

a formula of reduction.

Taken, between the limits 0 and $\frac{1}{2}\pi$,

$$\int_0^{\frac{1}{2}\pi}(\sin x)^n dx = \frac{n-1}{n}\int_0^{\frac{1}{2}\pi}(\sin x)^{n-2} dx,$$

$$\int_0^{\frac{1}{2}\pi}(\cos x)^n dx = \frac{n-1}{n}\int_0^{\frac{1}{2}\pi}(\cos x)^{n-2} dx.$$

First, suppose n even, $=2m$, then

$$\int_0^{\frac{1}{2}\pi} (\sin x)^{2m} dx = \frac{2m-1}{2m} \int_0^{\frac{1}{2}\pi} (\sin x)^{2m-2} dx$$

$$= \frac{2m-3}{2m-2} \frac{2m-1}{2m} \int_0^{\frac{1}{2}\pi} (\sin x)^{2m-4} dx,$$

and so on, finally being

$$= \frac{1 . 3 . 5 \dots 2m-1}{2 . 4 . 6 \dots 2m} \int_0^{\frac{1}{2}\pi} dx$$

$$= \frac{1 . 3 . 5 \dots 2m-1}{2 . 4 . 6 \dots 2m} \frac{\pi}{2};$$

and $\int_0^{\frac{1}{2}\pi} \cos^{2m} x \, dx$ has the same value.

Secondly, suppose n odd, $= 2m+1$, then, substituting as before in the formula of reduction, finally

$$\int_0^{\frac{1}{2}\pi} (\sin x)^{2m+1} dx = \frac{2 . 4 . 6 \dots 2m}{3 . 5 . 7 \dots 2m+1} \int_0^{\frac{1}{2}\pi} \sin x \, dx$$

$$= \frac{2 . 4 . 6 \dots 2m}{3 . 5 . 7 \dots 2m+1};$$

and $\int_0^{\frac{1}{2}\pi} (\cos x)^{2m+1} dx$ has the same value.

These are called *Wallis's Theorems*, and are of very great use in the Integral Calculus.

Examples.—Prove the following formulæ of reduction :

(i.) $\displaystyle\int x^n e^x dx = x^n e^x - n \int x^{n-1} e^x dx.$

(ii.) $\displaystyle\int (\log x)^n dx = x(\log x)^n - n \int (\log x)^{n-1} dx.$

(iii.) $\displaystyle\int x^m (\log x)^n dx = \frac{x^{m+1}(\log x)^n}{m+1} - \frac{n}{m+1}\int x^m (\log x)^{n-1} dx.$

(iv.) $\displaystyle\int (a^2 - x^2)^n dx = \frac{x(a^2-x^2)^n}{2n+1} + \frac{2na^2}{2n+1}\int (a^2-x^2)^{n-1} dx.$

(v.) $\displaystyle\int (a^2 + x^2)^{n-\frac{1}{2}} dx$
$$= \frac{x(a^2+x^2)^{n-\frac{1}{2}}}{2n} + \frac{2n-1}{2n}a^2 \int (a^2+x^2)^{n-\frac{3}{2}} dx.$$

(vi.) $\displaystyle\int x^m \sqrt{(a^2-x^2)}\,dx = -\frac{x^{m-1}(a^2-x^2)^{\frac{3}{2}}}{m+2}$
$$+ \frac{m-1}{m+2}a^2 \int x^{m-2}\sqrt{(a^2-x^2)}\,dx.$$

(vii.) $\displaystyle\int x^m \sqrt{(2ax-x^2)}\,dx = -\frac{x^{m-1}(2ax-x^2)^{\frac{3}{2}}}{m+2} -$
$$+ \frac{2m+1}{m+2}a \int x^{m-1}\sqrt{(2ax-x^2)}\,dx.$$

(viii.) $\displaystyle\int \frac{dx}{(Ax^2+2Bx+C)^n} = \frac{1}{2n-2}\,\frac{1}{AC-B^2}\,\frac{Ax+B}{(Ax^2+2Bx+C)^{n-1}}$
$$+ \frac{2n-3}{2n-2}\,\frac{A}{AC-B^2}\int \frac{dx}{(Ax^2+2Bx+C)^{n-1}}.$$

(ix.) $\displaystyle\int (\tan x)^n dx = \frac{(\tan x)^{n-1}}{n-1} - \int (\tan x)^{n-2} dx.$

(x.) $\displaystyle\int (\cot x)^n dx = -\frac{(\cot x)^{n-1}}{n-1} - \int (\cot x)^{n-2} dx.$

(xi.) $\displaystyle\int (\sec x)^n dx = \frac{(\sec x)^{n-2}\tan x}{n-1} + \frac{n-2}{n-1}\int (\sec x)^{n-2} dx.$

(xii.) $\int (\text{cosec } x)^n dx$

$$= -\frac{(\text{cosec } x)^{n-2}\cot x}{n-1} + \frac{n-2}{n-1}\int (\text{cosec } x)^{n-2} dx.$$

(xiii.) $\int (\text{vers } x)^n dx$

$$= -\frac{\sin x(\text{vers } x)^{n-1}}{n} + \frac{2n-1}{n}\int (\text{vers } x)^{n-1} dx.$$

(xiv.) $\int x^n \cos mx \, dx = \frac{1}{m} x^n \sin mx + \frac{n}{m^2} x^{n-1} \cos mx$

$$- \frac{n(n-1)}{m^2}\int x^{n-2} \cos mx \, dx.$$

(xv.) $\int e^{nx}(\cos x)^m dx = \frac{e^{nx}(\cos x)^{m-1}(n \cos x + m \sin x)}{m^2+n^2}$

$$+ \frac{m(m-1)}{m^2+n^2}\int e^{nx}(\cos x)^{m-2} dx.$$

(xvi.) $\int (\sin x)^m (\cos x)^n dx =$

$$\frac{(\sin x)^{m+1}(\cos x)^{n-1}}{m+n} + \frac{n-1}{m+n}\int (\sin x)^m (\cos x)^{n-2} dx,$$

or $\quad -\dfrac{(\sin x)^{m-1}(\cos x)^{n+1}}{m+n} + \dfrac{m-1}{m+n}\int (\sin x)^{m-2}(\cos x)^n dx.$

(xvii.) $\int_{0}^{\frac{1}{2}\pi} (\sin x)^{2m}(\cos x)^{2n} dx =$

$$\frac{1.3.5\ldots(2m-1).1.3.5\ldots(2n-1)}{2.4.6\ldots\ldots\ldots\ldots\ldots(2m+2n)}\frac{\pi}{2}.$$

(xviii.) Investigate formulæ of reduction for
$\int (\sin^{-1} x)^n dx, \quad \int (\text{vers}^{-1} x)^n dx, \quad \int (\sinh x)^n dx, \quad \int (\cosh x)^n dx,$
$\int (\tanh x)^n dx, \quad \int (\text{sech } x)^n dx, \quad \int (\sinh^{-1} x)^n dx, \quad \int (\cosh^{-1} x)^n dx,$
$\int \sin x^{\frac{1}{n}} dx, \quad \int \sinh x^{\frac{1}{n}} dx.$

68. *General Integration of Irrational Functions.*

The most general algebraical function of x which can be integrated by means of algebraical, circular, logarithmic or hyperbolic functions is of the form, or can be reduced to the form

$$\frac{A + B\sqrt{R}}{C + D\sqrt{R}},$$

where A, B, C, D are *rational integral* functions of x, and R is of the first or second degree in x, of the form $ax^2 + 2bx + c$.

(If R is of the third or fourth degree in x, *elliptic* functions are in general required in the integration; if R is of the fifth or higher degree, *hyperelliptic* functions are required.)

Rationalizing the denominator,

$$\frac{A + B\sqrt{R}}{C + D\sqrt{R}} = \frac{(A + B\sqrt{R})(C - D\sqrt{R})}{C^2 - D^2 R}$$

$$= \frac{AC - BDR}{C^2 - D^2 R} + \frac{BC - AD}{C^2 - D^2 R}\sqrt{R},$$

which is of the form $\dfrac{L}{M} + \dfrac{N}{M}\sqrt{R}$, where L, M, N are rational integral functions of x.

It has already been explained (§ 39) how the fraction $\dfrac{L}{M}$ can be integrated, so we must now consider the integration of $\dfrac{N}{M}\sqrt{R}$, or $\dfrac{NR}{M}\dfrac{1}{\sqrt{R}}$, as it will be written.

Three cases must be considered according to the form of $\sqrt{R}$, the irrational part.

Disregarding constant factors, R is either of the form

$$R = a^2 - (x-c)^2 \text{ or } (a-x)(x-\beta)\dots\dots\dots(1)$$

or

$$R = (x-c)^2 - a^2 \text{ or } (x-a)(x-\beta)\dots\dots\dots(2)$$

where $\qquad a = c + a, \quad \beta = c - a;$

so that $\qquad a + \beta = 2c, \quad a - \beta = 2a;$

or $\qquad R = a^2 + (x-c)^2 \dots\dots\dots\dots\dots\dots\dots\dots\dots(3)$

69. Form (1) is associated with the circle and ellipse (§ 54) and is reduced by putting

$$x - c = a \cos\theta;$$

or $\qquad x = a \cos^2 \tfrac{1}{2}\theta + \beta \sin^2 \tfrac{1}{2}\theta,$

or $\qquad \cos^2 \tfrac{1}{2}\theta = \dfrac{x - \beta}{a - \beta}, \ \sin^2 \tfrac{1}{2}\theta = \dfrac{a - x}{a - \beta}.$

Then $\qquad dx = -a \sin\theta\, d\theta,$

and $\qquad \sqrt{R} = a \sin\theta,$

so that $\qquad \dfrac{dx}{\sqrt{R}} = -d\theta.$

Then

$$\int_{\beta}^{x} \frac{dx}{\sqrt{(a-x \,.\, x-\beta)}} = 2 \sin^{-1}\sqrt{\frac{x-\beta}{a-\beta}} = 2 \cos^{-1}\sqrt{\frac{a-x}{a-\beta}};$$

$$\int_{x}^{a} \frac{dx}{\sqrt{(a-x \,.\, x-\beta)}} = 2 \sin^{-1}\sqrt{\frac{a-x}{a-\beta}} = 2 \cos^{-1}\sqrt{\frac{x-\beta}{a-\beta}}.$$

$$\int_{a}^{x} \frac{dx}{\sqrt{\{a^2 - (x-c)^2\}}} = \sin^{-1}\frac{x-c}{a}.$$

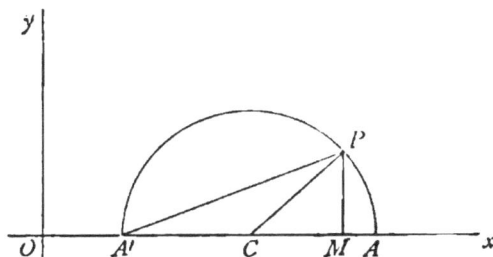

Fig. 19.

The geometrical interpretation is seen by reference to figure 19.

If $OA = a$, $OA' = \beta$, and a circle is described on AA' as diameter with centre C, then $OC = c$, and the radius $= a$.

If $OM = x$, then $ACP = \theta$, $AA'P = \frac{1}{2}\theta$, $AM = a - x$, $A'M = x - \beta$, and $MP^2 = (a - x)(x - \beta) = R$.

70. Form (2) is associated with the hyperbola (§ 56),

$$x^2 - y^2 = a^2 \, ;$$

and then, if the origin is changed from the centre a distance c to the left (fig. 20),

$$y^2 = (x - c)^2 - a^2 = R.$$

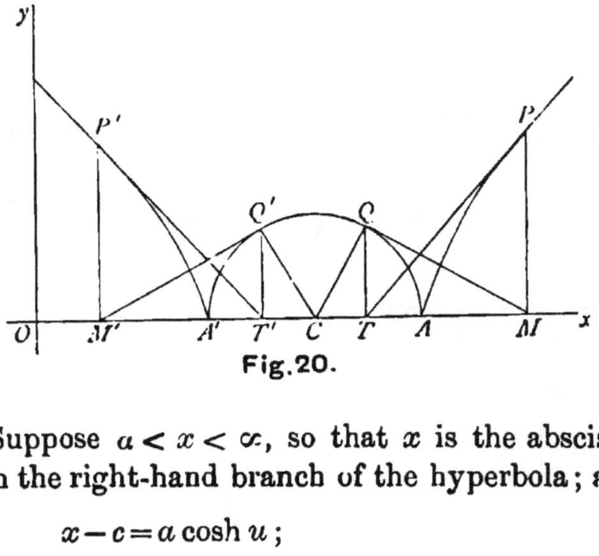

Fig. 20.

(i.) Suppose $a < x < \infty$, so that x is the abscissa of a point on the right-hand branch of the hyperbola; and put

$$x - c = a \cosh u \, ;$$

then
$$y = \sqrt{R} = a \sinh u,$$

and
$$\frac{dx}{\sqrt{R}} = du.$$

Also
$$x = c + a \cosh u = a \cosh^2 \tfrac{1}{2}u - \beta \sinh^2 \tfrac{1}{2}u,$$
$$x - a = (a - \beta) \sinh^2 \tfrac{1}{2}u, \quad x - \beta = (a - \beta) \cosh^2 \tfrac{1}{2}u.$$

Then
$$\frac{dx}{\sqrt{(x-a \cdot x - \beta)}} = du,$$

and
$$\int_a^x \frac{dx}{\sqrt{(x-a \cdot x - \beta)}} = 2 \sinh^{-1}\sqrt{\frac{x-a}{a-\beta}}$$

$$= 2 \cosh^{-1}\sqrt{\frac{x-\beta}{a-\beta}} = 2 \log \frac{\sqrt{(x-a)} + \sqrt{(x-\beta)}}{\sqrt{(a-\beta)}};$$

also
$$\int_{c+a}^x \frac{dx}{\sqrt{\{(x-c)^2 - a^2\}}} = \cosh^{-1}\frac{x-c}{a}.$$

(ii.) Suppose $\beta > x > -\infty$, so that x is the abscissa of a point on the left-hand branch; then we must put

$$x = c - a \cosh u' = -a \sinh^2\tfrac{1}{2}u' + \beta \cosh^2\tfrac{1}{2}u';$$

or $\quad a - x = (a - \beta)\cosh^2\tfrac{1}{2}u', \quad \beta - x = (a - \beta)\sinh^2\tfrac{1}{2}u';$

and then
$$\frac{dx}{\sqrt{(a-x \cdot \beta - x)}} = -du',$$

so that
$$\int_x^\beta \frac{dx}{\sqrt{(a-x \cdot \beta - x)}} = 2 \cosh^{-1}\sqrt{\frac{a-x}{a-\beta}}$$

$$= 2 \sinh^{-1}\sqrt{\frac{\beta-x}{a-\beta}} = 2 \log \frac{\sqrt{(a-x)} + \sqrt{(\beta-x)}}{\sqrt{(a-\beta)}};$$

also
$$\int_x^{c-a} \frac{dx}{\sqrt{\{(c-x)^2 - a^2\}}} = \cosh^{-1}\frac{c-x}{a}.$$

Referring to fig. 20, the sectorial area $CAP = \tfrac{1}{2}a^2u$, and $\qquad CM = a \cosh u, \quad MP = a \sinh u$; also $CA'P' = \tfrac{1}{2}a^2u', \quad CM' = a \cosh u', \quad M'P' = a \sinh u'$.

The circular functions can be employed in the reduction of this form by putting

$$x - c = a \sec \theta, \quad \text{and} \quad y = a \tan \theta;$$

and then $ACQ = \theta$, where MQ is the tangent to the circle.

Then $\quad \displaystyle\int \frac{dx}{\sqrt{R}} = \int \sec \theta \, d\theta = \cosh^{-1}\sec \theta = \cosh^{-1}\frac{x-c}{a}.$

71. Form (3) is associated with the conjugate hyperbola (§ 57),

$$y^2 - (x-c)^2 = a^2,$$

and is reduced by putting

$$x - c = a \sinh v,$$

and
$$y = \sqrt{R} = a \cosh v \,;$$

so that
$$\frac{dx}{\sqrt{R}} = dv,$$

and
$$\int_c^x \frac{dx}{\sqrt{\{a^2 + (x-c)^2\}}} = \sinh^{-1} \frac{x-c}{a}$$

$$= \log \frac{\sqrt{\{a^2 + (x-c)^2\}} + x - c}{a}.$$

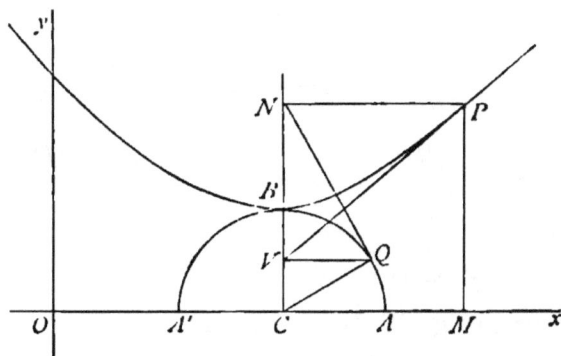

Fig.21 .

Referring to fig 21, the sectorial area $BCP = \frac{1}{2}a^2 v$,

and $NP = x - c = a \sinh v, \; MP = y = a \cosh v.$

In reducing by means of the circular functions, put

$$x - c = a \tan \phi,$$
then
$$y = \sqrt{R} = a \sec \phi$$

and the angle $BCQ = \phi$, where NQ is the tangent to the circle.

Then

$$\int_c^z \frac{dx}{\sqrt{\{a^2 + (x-c)^2\}}} = \int_0^\phi \sec \phi \, d\phi = \cosh^{-1} \frac{\sqrt{\{a^2 + (x-c)^2\}}}{a}.$$

72. Now to integrate $\frac{NR}{M} \frac{1}{\sqrt{R}}$, the rational part $\frac{NR}{M}$ must be resolved into a quotient and partial fractions in the manner already explained (§ 39); and then for any term Cx^n of the quotient the corresponding integral $C \int \frac{x^n dx}{\sqrt{R}}$ is made to depend upon the integration of integral powers of $\cos \theta$, $\sin \theta$, $\cosh u$ or $\sinh u$ by the substitutions of §§ 69-71, according to the form of R.

The integral $\int \frac{dx}{(x-p)\sqrt{R}}$, corresponding to the partial fraction $\frac{A}{x-p}$ of $\frac{NR}{M}$ (§ 39), and generally the integrals of the form $\int \frac{dx}{(x-p)^n \sqrt{R}}$, corresponding to the partial fractions $\frac{B_n}{(x-p)^n}$ of $\frac{NR}{M}$ (§ 41) are made to depend upon the integrals of the form $\int \frac{z^m dz}{\sqrt{R'}}$, where R' is of the form $a'z^2 + 2b'z + c'$, by the substitution $x - p = \frac{1}{z}$.

With the substitutions of §§ 69-71, the integral $\int \frac{dx}{(x-p)^n \sqrt{R}}$ is made to depend upon integrals of the form $\int \frac{d\theta}{(a + \beta \cos \theta)^n}$ or $\int \frac{du}{(a + \beta \cosh u)^n}$ or $\int \frac{dv}{(a + \beta \sinh v)^n}$.

Lastly, integrals of the form

$$\int \frac{Px+Q}{(Ax^2+2Bx+C)\sqrt{R}},$$

corresponding to partial fractions of the form (§ 40)

$$\frac{Px+Q}{Ax^2+2Bx+C},$$

or, more generally, integrals of the form

$$\int \frac{Px+Q}{(Ax^2+2Bx+C)^n\sqrt{R}},$$

are reduced by the substitutions of §§ 69-71 to integrals of the form

$$\int \frac{P'\cos\theta+Q'}{(A'\cos^2\theta+2B'\cos\theta+C')^n}\,d\theta,$$

or

$$\int \frac{P'\cosh u+Q'}{(A'\cosh^2u+2B'\cosh u+C')^n}\,du,$$

or

$$\int \frac{P'\sinh v+Q'}{(A'\sinh^2v+2B'\sinh v+C')^n}\,dv;$$

and these again are reduced to integrals of rational fractions, as in fact $\int \frac{NR}{M}\,\frac{dx}{\sqrt{R}}$ is reduced, by the substitution of z for $\tan\frac{1}{2}\theta$, or $\tanh\frac{1}{2}u$, or $\tanh\frac{1}{2}v$.

The integration of irrational functions, which can be effected by means of algebraical, circular, or hyperbolic functions, can thus always be reduced to the integration of rational functions, by means of the above substitutions, and the integrals expressed by algebraical functions, or by the circular and hyperbolic tangents and their inverse functions; these are therefore the only *transcendental* functions we need employ, but as mentioned before (§ 24) the other circular and hyperbolic functions are employed to avoid irrational and complicated expressions.

Elliptic and hyperelliptic integrals cannot be reduced by any of the previous substitutions to the integrals of rational algebraical functions.

73. Consider more particularly the integral

$$\int \frac{Px+Q}{(Ax^2+2Bx+C)\sqrt{(ax^2+2bx+c)}}dx,$$

where $Ax^2+2Bx+C$ is always positive, and does not split up into real linear factors, and therefore $AC-B^2$ is positive.

By means of the substitution

$$\frac{ax^2+2bx+c}{Ax^2+2Bx+C}=y^2,$$

the result will be found to be of the form

$$L\sin^{-1}\left(k\frac{ax^2+2bx+c}{Ax^2+2Bx+C}\right)^{\frac{1}{2}}+M\sinh^{-1}\left(-k'\frac{ax^2+2bx+c}{Ax^2+2Bx+C}\right)^{\frac{1}{2}}\text{(i.)}$$

if $ac-b^2$ is negative ; or

$$L\cos^{-1}\left(k\frac{ax^2+2bx+c}{Ax^2+2Bx+C}\right)^{\frac{1}{2}}+M\cosh^{-1}\left(k'\frac{ax^2+2bx+c}{Ax^2+2Bx+C}\right)^{\frac{1}{2}}\text{(ii.)}$$

if $ac-b^2$ is positive.

Here L and M are constants to be determined, most conveniently by differentiating (i.) and (ii.); and k, k' are the values of λ, which make

$$Ax^2+2Bx+C-\lambda(ax^2+2bx+c)$$

a perfect square, and therefore

$$(A-\lambda a)(C-\lambda c)-(B-\lambda b)^2=0.$$

Denoting $Ax^2 + 2Bx + C$ by U, and $ax^2 + 2bx + c$ by V;
also denoting $\dfrac{dV}{dx}U - V\dfrac{dU}{dx}$ by $2J$, then $4J$ is called the Jacobian of U and V (Salmon, *Higher Algebra*).

Case (i.) If $ac - b^2$ is negative, and
$$U - kV = (mx + n)^2, \quad U - k'V = (m'x + n')^2,$$
then $\qquad J = (b^2 - ac)^{\frac{1}{2}}(mx + n)(m'x + n').$

Then differentiating (i.),
$$\frac{Lk^{\frac{1}{2}}J}{UV^{\frac{1}{2}}(mx + n)} + \frac{M(-k')^{\frac{1}{2}}J}{UV^{\frac{1}{2}}(m'x + n')} = \frac{Px + Q}{UV^{\frac{1}{2}}},$$

or $\qquad Lk^{\frac{1}{2}}(m'x + n') + M(-k')^{\frac{1}{2}}(mx + n) = \dfrac{Px + Q}{(b^2 - ac)^{\frac{1}{2}}};$

an identical equation, so that putting $mx + n = 0$,
$$L = \frac{Pn - Qm}{k^{\frac{1}{2}}(m'n - mn')(b^2 - ac)^{\frac{1}{2}}};$$

and putting $m'x + n' = 0$,
$$M = \frac{Pn' - Qm'}{(-k')^{\frac{1}{2}}(mn' - m'n)(b^2 - ac)^{\frac{1}{2}}}.$$

Case (ii.) If $ac - b^2$ is positive, and
$$U - kV = (mx + n)^2, \quad U - k'V = -(m'x + n')^2,$$
then $\qquad J = (ac - b^2)^{\frac{1}{2}}(mx + n)(m'x + n');$
and L and M are determined as before.

Similarly, $\displaystyle\int \frac{P\cos\theta + Q}{A\cos^2\theta + 2B\cos\theta + C}d\theta$

$$= L\sin^{-1}\left(\frac{k\sin^2\theta}{A\cos^2\theta + 2B\cos\theta + C}\right)^{\frac{1}{2}}$$

$$+ M\sinh^{-1}\left(\frac{-k'\sin^2\theta}{A\cos^2\theta + 2B\cos\theta + C}\right)^{\frac{1}{2}};$$

$$\int \frac{P\cosh u + Q}{A\cosh^2 u + 2B\cosh u + C}du$$

$$= L\sin^{-1}\left(\frac{k\sinh^2 u}{A\cosh^2 u + 2B\cosh u + C}\right)^{\frac{1}{2}}$$

$$+ M\sinh^{-1}\left(\frac{-k'\sinh^2 u}{A\cosh^2 u + 2B\cosh u + C}\right)^{\frac{1}{2}};$$

and

$$\int \frac{P\sinh v + Q}{A\sinh^2 v + 2B\sinh v + C}dv$$

$$= L\cos^{-1}\left(\frac{k\cosh^2 v}{A\sinh^2 v + 2B\sinh v + C}\right)^{\frac{1}{2}}$$

$$+ M\cosh^{-1}\left(\frac{k'\cosh^2 v}{A\sinh^2 v + 2B\sinh v + C}\right)^{\frac{1}{2}}.$$

Then

$$\int_0^\pi \frac{P\cos\theta + Q}{A\cos^2\theta + 2B\cos\theta + C}d\theta = \pi L;$$

$$\int_0^\infty \frac{P\cosh u + Q}{A\cosh^2 u + 2B\cosh u + C}du$$

$$= L\sin^{-1}\left(\frac{k}{A}\right)^{\frac{1}{2}} + M\sinh^{-1}\left(-\frac{k'}{A}\right)^{\frac{1}{2}};$$

$$\int_{-\infty}^\infty \frac{P\sinh v + Q}{A\sinh^2 v + 2B\sinh v + C}dv = 2L\cos^{-1}\left(\frac{k}{A}\right)^{\frac{1}{2}}.$$

The result of the integration of

$$\int \frac{Px + Q}{(Ax^2 + 2Bx + C)^n \sqrt{(ax^2 + 2bx + c)}}dx$$

etc., can be inferred by differentiating the above results $n-1$ times with respect to C, treating C for the moment as a variable quantity.

74. *General Integration of Circular and Hyperbolic Functions.*

A large class of the preceding integrals are seen to depend on

$$\int \frac{dx}{(a+\beta \cos x)^n} \text{ or } \int \frac{dx}{(a+\beta \cosh x)^n},$$

and we shall employ a general method of reducing them, depending on geometrical considerations.

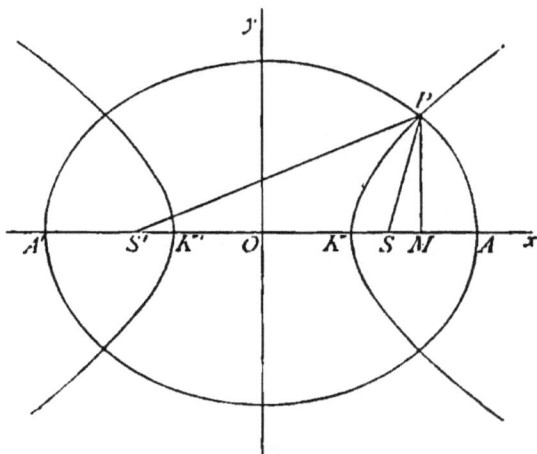

Fig. 22

Using the variables ξ and η explained in § 29 for confocal conics, and denoting SP by r, $S'P$ by r', the angle xSP by θ and the angle $xS'P$ by θ' (fig. 22); then in Astronomy θ or θ' is called the *true anomaly* of P.

If P moves on an ellipse for which η is constant, then ξ is called the *eccentric anomaly*; if P moves on a confocal hyperbola, ξ is constant and η varies.

From § 29,

$$r = c(\cosh \eta - \cos \xi), \ r' = c(\cosh \eta + \cos \xi),$$

so that $SP = AK, \ S'P = KA'$; and

$$\cos\theta = \frac{x-c}{r} = \frac{\cos\xi\cosh\eta - 1}{\cosh\eta - \cos\xi}, \quad \cos\theta' = \frac{x+c}{r'} = \frac{\cos\xi\cosh\eta + 1}{\cosh\eta + \cos\xi}.$$

Therefore

$$\sin\theta = \frac{\sin\xi\sinh\eta}{\cosh\eta - \cos\xi}, \quad \sin\theta' = \frac{\sin\xi\sinh\eta}{\cosh\eta + \cos\xi};$$

and $\tan\tfrac{1}{2}\theta = \tan\tfrac{1}{2}\xi\coth\tfrac{1}{2}\eta, \quad \tan\tfrac{1}{2}\theta' = \tan\tfrac{1}{2}\xi\tanh\tfrac{1}{2}\eta.$

Differentiating the last two equations logarithmically (§ 35),

$$\frac{d\theta}{\sin\theta} = \frac{d\xi}{\sin\xi} = -\frac{d\eta}{\sinh\eta};$$

$$\frac{d\theta'}{\sin\theta'} = \frac{d\xi}{\sin\xi} = \frac{d\eta}{\sinh\eta}.$$

Therefore

$$d\theta = \frac{\sinh\eta\, d\xi}{\cosh\eta - \cos\xi} = -\frac{\sin\xi\, d\eta}{\cosh\eta - \cos\xi};$$

$$d\theta' = \frac{\sinh\eta\, d\xi}{\cosh\eta + \cos\xi} = \frac{\sin\xi\, d\eta}{\cosh\eta + \cos\xi}.$$

But

$$\cosh\eta + \cos\theta = \frac{\sinh^2\eta}{\cosh\eta - \cos\xi}, \quad \cos\xi - \cos\theta = \frac{\sin^2\xi}{\cosh\eta - \cos\xi};$$

$$\cosh\eta - \cos\theta' = \frac{\sinh^2\eta}{\cosh\eta + \cos\xi}, \quad \cos\theta' - \cos\xi = \frac{\sin^2\xi}{\cosh\eta + \cos\xi}.$$

Therefore

$$(\cosh\eta + \cos\theta)^{n-1}d\theta = \frac{(\sinh\eta)^{2n-1}d\xi}{(\cosh\eta - \cos\xi)^n},$$

and integrating,

$$\int_0^\xi \frac{(\sinh \eta)^{2n-1} d\xi}{(\cosh \eta - \cos \xi)^n} = \int_0^\theta (\cosh \eta + \cos \theta)^{n-1} d\theta \dots\dots(1)$$

Similarly,

$$\int_0^\eta \frac{(\sin \xi)^{2n-1} d\eta}{(\cosh \eta - \cos \xi)^n} = \int_\theta^\pi (\cos \xi - \cos \theta)^{n-1} d\theta \dots\dots(2)$$

$$\int_0^\xi \frac{(\sinh \eta)^{2n-1} d\xi}{(\cosh \eta + \cos \xi)^n} = \int_0^{\theta'} (\cosh \eta - \cos \theta')^{n-1} d\theta' \dots\dots(3)$$

$$\int_0^\eta \frac{(\sin \xi)^{2n-1} d\eta}{(\cosh \eta + \cos \xi)^n} = \int_0^{\theta'} (\cos \theta' - \cos \xi)^{n-1} d\theta' \dots\dots(4)$$

Here the denominators do not vanish, and the integrals do not become infinite, and can be expressed by means of circular functions.

But

$$\sec \theta = \frac{\cosh \eta - \cos \xi}{\cos \xi \cosh \eta - 1}, \quad \sec \theta' = \frac{\cosh \eta + \cos \xi}{\cos \xi \cosh \eta + 1};$$

therefore

$$\int_0^\xi \frac{(\sinh \eta)^{2n-1} d\xi}{(\cos \xi \cosh \eta - 1)^n} = \int_0^\theta (\cosh \eta + \cos \theta)^{n-1} (\sec \theta)^n d\theta \dots(5)$$

$$\int_0^\eta \frac{(\sin \xi)^{2n-1} d\eta}{(\cos \xi \cosh \eta - 1)^n} = \int_\theta^\pi (\cos \xi - \cos \theta)^{n-1} (\sec \theta)^n d\theta \dots(6)$$

$$\int_0^\xi \frac{(\sinh \eta)^{2n-1} d\xi}{(\cos \xi \cosh \eta + 1)^n} = \int_0^{\theta'} (\cosh \eta - \cos \theta')^{n-1} (\sec \theta')^n d\theta' \dots(7)$$

$$\int_0^\eta \frac{(\sin \xi)^{2n-1} d\eta}{(\cos \xi \cosh \eta + 1)^n} = \int_0^{\theta'} (\cos \theta' - \cos \xi)^{n-1} (\sec \theta')^n d\theta' \dots(8)$$

Here the denominators can vanish, and the integrals become infinite, so that they require hyperbolic functions for their expression.

According to the relative values of α and β, the integrals mentioned at the beginning of this article will assume one of the forms (1) to (8).

For $n = 1$,

$$\int_0^\xi \frac{\sinh \eta \, d\xi}{\cosh \eta - \cos \xi} = \theta = \cos^{-1}\frac{\cos \xi \cosh \eta - 1}{\cosh \eta - \cos \xi},$$

$$\int_0^\eta \frac{\sin \xi \, d\eta}{\cosh \eta - \cos \xi} = \pi - \theta = \pi - \cos^{-1}\frac{\cos \xi \cosh \eta - 1}{\cosh \eta - \cos \xi},$$

$$\int_0^\xi \frac{\sinh \eta \, d\xi}{\cosh \eta + \cos \xi} = \int_0^\eta \frac{\sin \xi \, d\eta}{\cosh \eta + \cos \xi} = \theta' = \cos^{-1}\frac{\cos \xi \cosh \eta + 1}{\cosh \eta + \cos \xi},$$

$$\int_0^\xi \frac{\sinh \eta \, d\xi}{\cos \xi \cosh \eta - 1} = \int_0^\theta \sec \theta \, d\theta = \cosh^{-1}\sec \theta$$

$$= \cosh^{-1}\frac{\cosh \eta - \cos \xi}{\cos \xi \cosh \eta - 1},$$

$$\int_0^\eta \frac{\sin \xi \, d\eta}{\cos \xi \cosh \eta - 1} = \int_\theta^\pi \sec \theta \, d\theta = \cosh^{-1}(-\sec \theta)$$

$$= \cosh^{-1}\frac{\cos \xi - \cosh \eta}{\cos \xi \cosh \eta - 1},$$

$$\int_0^\xi \frac{\sinh \eta \, d\xi}{\cos \xi \cosh \eta + 1} = \int_0^\eta \frac{\sin \xi \, d\eta}{\cos \xi \cosh \eta + 1} = \int_0^{\theta'} \sec \theta' \, d\theta'$$

$$= \cosh^{-1}\sec \theta' = \cosh^{-1}\frac{\cosh \eta + \cos \xi}{\cos \xi \cosh \eta + 1}.$$

The geometrical interpretation of the reduction of the integral

$$\int \frac{dx}{(\alpha + \beta \sinh x)^n}$$

is left as an exercise for the student.

General Examples of Integration.

1. $\displaystyle\int\frac{dx}{ax^2+2bx+c}=-\frac{1}{\sqrt{(ac-b^2)}}\tan^{-1}\frac{ax+b}{\sqrt{(ac-b^2)}}$, or

$-\dfrac{1}{\sqrt{(b^2-ac)}}\tanh^{-1}\dfrac{ax+b}{\sqrt{(b^2-ac)}}$, or $-\dfrac{1}{\sqrt{(b^2-ac)}}\coth^{-1}\dfrac{ax+b}{\sqrt{(b^2-ac)}}$;

the real form to be taken.

Express the results also as functions of $ax^2+2bx+c$.

2. $\displaystyle\int\frac{dx}{\sqrt{(ax^2+2bx+c)}}=$

$\dfrac{1}{\sqrt{a}}\sinh^{-1}\dfrac{ax+b}{\sqrt{(ac-b^2)}}=\dfrac{1}{\sqrt{a}}\cosh^{-1}\sqrt{\left(a\dfrac{ax^2+2bx+c}{ac-b^2}\right)}$,

or $\dfrac{1}{\sqrt{a}}\cosh^{-1}\dfrac{ax+b}{\sqrt{(b^2-ac)}}=\dfrac{1}{\sqrt{a}}\sinh^{-1}\sqrt{\left(a\dfrac{ax^2+2bx+c}{b^2-ac}\right)}$,

or

$\dfrac{1}{\sqrt{(-a)}}\cos^{-1}\dfrac{ax+b}{\sqrt{(b^2-ac)}}=\dfrac{1}{\sqrt{(-a)}}\sin^{-1}\sqrt{\left(-a\dfrac{ax^2+2bx+c}{b^2-ac}\right)}$.

3. $\displaystyle\int\frac{dx}{(a-x)\sqrt{(x-b)}}=\frac{2}{\sqrt{(a-b)}}\sinh^{-1}\sqrt{\frac{x-b}{a-x}}$,

or $\dfrac{2}{\sqrt{(a-b)}}\cosh^{-1}\sqrt{\dfrac{x-b}{x-a}}$, or $\dfrac{2}{\sqrt{(b-a)}}\cos^{-1}\sqrt{\dfrac{x-b}{x-a}}$.

4. $\displaystyle\int\frac{dx}{(a-x)\sqrt{(b-x)}}=\frac{2}{\sqrt{(b-a)}}\sinh^{-1}\sqrt{\frac{b-x}{x-a}}$,

or $\dfrac{2}{\sqrt{(b-a)}}\cosh^{-1}\sqrt{\dfrac{b-x}{a-x}}$, or $\dfrac{2}{\sqrt{(a-b)}}\cos^{-1}\sqrt{\dfrac{b-x}{a-x}}$.

5. $\displaystyle\int\frac{dx}{(x-p)\sqrt{(ax^2+2bx+c)}}=$

$-\dfrac{1}{\sqrt{(ap^2+2bp+c)}}\sinh^{-1}\dfrac{(ap+b)x+bp+c}{(x-p)\sqrt{(ac-b^2)}}$

$=-\dfrac{1}{\sqrt{(ap^2+2bp+c)}}\cosh^{-1}\dfrac{\sqrt{(ax^2+2bx+c)}\sqrt{(ap^2+2bp+c)}}{(x-p)\sqrt{(ac-b^2)}}$,

or
$$-\frac{1}{\sqrt{(ap^2+2bp+c)}}\cosh^{-1}\frac{(ap+b)x+bp+c}{(x-p)\sqrt{(b^2-ac)}}$$

$$=-\frac{1}{\sqrt{(ap^2+2bp+c)}}\sinh^{-1}\frac{\sqrt{(ax^2+2bx+c)}\sqrt{(ap^2+2bp+c)}}{(x-p)\sqrt{(b^2-ac)}},$$

or
$$\frac{1}{\sqrt{(-ap^2-2bp-c)}}\sin^{-1}\frac{(ap+b)x+bp+c}{(x-p)\sqrt{(b^2-ac)}}$$

$$=\frac{1}{\sqrt{(-ap^2-2bp-c)}}\cos^{-1}\frac{\sqrt{(ax^2+2bx+c)}\sqrt{(-ap^2-2bp-c)}}{(x-p)\sqrt{(b^2-ac)}}.$$

6. $\displaystyle\int\frac{Px+Q}{(Ax^2+C)\sqrt{(ax^2+c)}}dx=$

$$-\frac{P}{\sqrt{A}\sqrt{(Ac-aC)}}\cosh^{-1}\sqrt{\left(\frac{A}{a}\frac{ax^2+c}{Ax^2+C}\right)}$$

$$+\frac{Q}{\sqrt{C}\sqrt{(Ac-aC)}}\cos^{-1}\sqrt{\left(\frac{C}{c}\frac{ax^2+c}{Ax^2+C}\right)},$$

or
$$-\frac{P}{\sqrt{A}\sqrt{(Ac-aC)}}\sinh^{-1}\sqrt{\left(-\frac{A}{a}\frac{ax^2+c}{Ax^2+C}\right)}$$

$$-\frac{Q}{\sqrt{C}\sqrt{(Ac-aC)}}\sin^{-1}\sqrt{\left(\frac{C}{c}\frac{ax^2+c}{Ax^2+C}\right)},$$

or
$$-\frac{P}{\sqrt{A}\sqrt{(aC-Ac)}}\cos^{-1}\sqrt{\left(\frac{A}{a}\frac{ax^2+c}{Ax^2+C}\right)}$$

$$+\frac{Q}{\sqrt{C}\sqrt{(aC-Ac)}}\cosh^{-1}\sqrt{\left(\frac{C}{c}\frac{ax^2+c}{Ax^2+C}\right)},$$

or
$$\frac{P}{\sqrt{A}\sqrt{(aC-Ac)}}\sin^{-1}\sqrt{\left(\frac{A}{a}\frac{ax^2+c}{Ax^2+C}\right)}$$

$$+\frac{Q}{\sqrt{C}\sqrt{(aC-Ac)}}\sinh^{-1}\sqrt{\left(-\frac{C}{c}\frac{ax^2+c}{Ax^2+C}\right)}.$$

7. $\displaystyle\int_0^\pi \frac{P\cos\theta + Q}{\cos^2\theta - 2\sin a\cos\theta + 1}d\theta =$

$$\pi\frac{P\sin\tfrac{1}{2}a + Q\cos\tfrac{1}{2}a}{\sqrt{(2\cos^3 a)}}, \text{ if } \cos a \text{ is positive,}$$

or

$$\pi\frac{P\sin\tfrac{1}{2}a + Q\cos\tfrac{1}{2}a}{\sqrt{(-2\cos^3 a)}}, \text{ if } \cos a \text{ is negative.}$$

8. $\displaystyle\int_0^\pi \frac{P\cos\theta + Q}{a^2\cos^2\theta - 2ac\cos a\cos\theta + c^2}d\theta$

$$= \pi\frac{P\sqrt{(a^2+c^2-m^2)} + Q\sqrt{(a^2+c^2+m^2)}}{m^2\sqrt{(c^2-a^2+m^2)}},$$

where $\qquad m^4 = a^4 - 2a^2c^2\cos 2a + c^4.$

9. $\displaystyle\int_0 \frac{x^{2m}dx}{1 - x^{2n}} =$

$$\frac{1}{2n}\sum_{r=1}^{r=n}\left\{ \sin\left(\frac{2m+1}{n}r\pi\right)\tan^{-1}\left(\frac{2x\sin\dfrac{r\pi}{n}}{1-x^2}\right)\right.$$

$$\left. +\cos\left(\frac{2m+1}{n}r\pi\right)\tanh^{-1}\left(\frac{2x\cos\dfrac{r\pi}{n}}{1+x^2}\right)\right\}.$$

10. $\displaystyle\int_0 \frac{x^{2m}dx}{1 + x^{2n}} =$

$$\frac{1}{2n}\sum_{r=1}^{r=n}\left\{ \sin\left\{\frac{2m+1}{2n}(2r-1)\pi\right\}\tan^{-1}\left(\frac{2x\sin\dfrac{2r-1}{2n}\pi}{1-x^2}\right)\right.$$

$$\left. +\cos\left\{\frac{2m+1}{2n}(2r-1)\pi\right\}\tanh^{-1}\left(\frac{2x\cos\dfrac{2r-1}{2n}\pi}{1+x^2}\right)\right\}.$$

11. Find $\displaystyle\int(\tan x)^{\frac{m}{n}}dx$ and $\displaystyle\int(\tanh x)^{\frac{m}{n}}dx$.

(Put $\tan x$ or $\tanh x = y^n$).

12. $\displaystyle\int\sqrt{(2\cot x)}dx = \sinh^{-1}\sqrt{(\sin 2x)} + \sin^{-1}\sqrt{(\sin 2x)}.$

13. $\displaystyle\int\sqrt{(\coth x)}dx = \tanh^{-1}\sqrt{(\tanh x)} + \tan^{-1}\sqrt{(\tanh x)}.$

14. $\int \dfrac{x^2-1}{x} \dfrac{dx}{\sqrt{(x^4+cx^2+1)}} = $ (putting $x+\dfrac{1}{x}=y$)

$\qquad \sinh^{-1}\dfrac{x^2+1}{x\sqrt{(c-2)}}$ or $\cosh^{-1}\dfrac{x^2+1}{x\sqrt{(2-c)}}.$

15. $\int \dfrac{x^2+1}{x} \dfrac{dx}{\sqrt{(x^4-cx^2+1)}} = $ (putting $x-\dfrac{1}{x}=y$)

$\qquad \cosh^{-1}\dfrac{x^2-1}{x\sqrt{(c-2)}}$ or $\sinh^{-1}\dfrac{x^2-1}{x\sqrt{(2-c)}}.$

16. $\int \dfrac{1-x^2}{1+x^2} \dfrac{dx}{\sqrt{(1+cx^2+x^4)}} =$

$\qquad \dfrac{1}{\sqrt{(c-2)}}\sinh^{-1}\dfrac{x\sqrt{(c-2)}}{1+x^2}$ or $-\dfrac{1}{\sqrt{(2-c)}}\sin^{-1}\dfrac{x\sqrt{(2-c)}}{1+x^2}.$

17. $\int \dfrac{1+x^2}{1-x^2} \dfrac{dx}{\sqrt{(1-cx^2+x^4)}} =$

$\qquad \dfrac{1}{\sqrt{(c-2)}}\sin^{-1}\dfrac{x\sqrt{(c-2)}}{1-x^2}$ or $\dfrac{1}{\sqrt{(2-c)}}\sinh^{-1}\dfrac{x\sqrt{(2-c)}}{1-x^2}.$

18. $\int \dfrac{x^2-1}{x} \dfrac{dx}{\sqrt{R}} = \dfrac{1}{\sqrt{a}} \sinh^{-1}\dfrac{ax^2+bx+a}{x\sqrt{\Delta}}$

$\qquad = \dfrac{1}{\sqrt{a}} \cosh^{-1}\dfrac{\sqrt{a}\sqrt{R}}{x\sqrt{\Delta}},$ etc. ;

where $\quad R=ax^4+2bx^3+cx^2+2bx+a,\ \Delta=ac-2a^2-b^2.$

(Put $x+\dfrac{1}{x}=y$).

Determine also $\quad \int \dfrac{x^2+1}{x} \dfrac{dx}{\sqrt{R'}},$

where $\quad R'=ax^4-2bx^3+cx^2+2bx+a.$ (Put $x-\dfrac{1}{x}=y$).

19. Determine the conditions in order that $\int \dfrac{M}{N}\sqrt{R}dx$ (§ 72) can be expressed by a single term of the form $\tan^{-1}\dfrac{q}{p}\sqrt{R}$, or $\tanh^{-1}\dfrac{q}{p}\sqrt{R}$, where $M,\ N,\ p,\ q,\ R$ are rational integral functions of x. (Abel, *Œuvres*, I. p. 104).

20. Prove that the co-ordinates of the centroid of the arc of the curve $y = a \log \sec\dfrac{x}{a}$, measured from $x=0$, are

$$x - y \cot\frac{x}{a}, \quad y - a + x \cot\frac{x}{a},$$

supposing the line density varies as $\cosh\dfrac{s}{a}$.

21. Determine the co-ordinates of the centroid of the arc and of the area of the catenary $y = a\cosh\dfrac{x}{a}$, measured from $x=0$, and prove that for the figure formed by the revolution of the catenary round the axis of x,
$$S = \pi(ax + sy), \quad V = \tfrac{1}{2}aS.$$

22 Prove that for the figure formed by the revolution of the tractrix (ex. 92, p. 58) round the axis of x,
$$S = 2\pi a(a - y), \quad V = \tfrac{1}{3}\pi(a^2 - y^2)^{\frac{3}{2}}.$$

23. Rectify the *cissoid* $y^2(2a - x) = x^3$, and determine also A, S and V, in terms of x, starting from $x=0$.

24. Prove that, if $y = \dfrac{x^3 - a^3}{3x^2}$,

$$\int_y^\infty \frac{dy}{\sqrt{(4y^3 + a^3)}} = \sqrt{3}\int_x^\infty \frac{dx}{\sqrt{(4x^3 - a^3)}}.$$

25. Prove that, if
$$(A + 3Bx + 3Cx^2 + Dx^3)(A + 3By + 3Cy^2 + Dy^3)(A + 3Bz + 3Cz^2 + Dz^3)$$
$$= \{A + B(x + y + z) + C(yz + zx + xy) + Dxyz^3\},$$
then
$$\int_0^{} (A + 3Bx + 3Cx^2 + Dx^3)^{-\frac{2}{3}}dx + \int_0^{} (A + 3By + 3Cy^2 + Dy^3)^{-\frac{2}{3}}dy$$
$$+ \int_0^{} (A + 3Bz + 3Cz^2 + Dz^3)^{-\frac{2}{3}}dz = 0.$$

(*Quarterly Journal of Math.*, xix., p. 158; xx., p. 179.)

CHAPTER III.

SUCCESSIVE DIFFERENTIATION.

75. The operation of differentiation with respect to x being represented by the symbol $\dfrac{d}{dx}$, a second differentiation is represented by $\dfrac{d}{dx}\dfrac{d}{dx}$, which is written $\left(\dfrac{d}{dx}\right)^2$ or $\dfrac{d^2}{dx^2}$; and generally the operation of differentiating n times is represented by $\left(\dfrac{d}{dx}\right)^n$ or $\dfrac{d^n}{dx^n}$, so that the nth d.c. of y with respect to x will be written $\left(\dfrac{d}{dx}\right)^n y$ or $\dfrac{d^n y}{dx^n}$.

If $y = fx$, and $\dfrac{dy}{dx}$ is denoted by $f'x$, then $\dfrac{d^2 y}{dx^2}$ is denoted by $f''x$, and generally $\dfrac{d^n y}{dx^n}$ by $f^n x$.

The difference between $\left(\dfrac{dy}{dx}\right)^n$ and $\dfrac{d^n y}{dx^n}$ must be carefully observed; $\left(\dfrac{dy}{dx}\right)^n$ meaning the nth power of $\dfrac{dy}{dx}$, but $\dfrac{d^n y}{dx^n}$ meaning the nth d.c. of y.

Hitherto $\dfrac{dy^2}{dx^2}$ has generally been used for $\left(\dfrac{dy}{dx}\right)^2$.

I

The successive d.c.'s of a function are required, among other purposes, in the expansion of a function by Taylor's Theorem, as explained in the next chapter.

76. For instance (§ 3),

$$\frac{dx^m}{dx} = mx^{m-1};$$

therefore $\quad \dfrac{d^2x^m}{dx^2} = m(m-1)x^{m-2},$

$$\frac{d^3x^m}{dx^3} = m(m-1)(m-2)x^{m-3},$$

and generally

$$\frac{d^nx^m}{dx^n} = m(m-1)(m-2)\ldots(m-n+1)x^{m-n}.$$

If m is a positive integer

$$\frac{d^mx^m}{dx^m} = m(m-1)(m-2)\ldots2\,.\,1, \text{ denoted by } m\,!,$$

and all the higher d.c.'s vanish.

Generally to differentiate successively any rational function of x with respect to x, the function should first be resolved into its partial fractions (§ 39), and then the nth d.c. of a partial fraction

$$\frac{A}{x-a} \text{ will be } A\frac{(-1)^n n\,!}{(x-a)^{n+1}},$$

and of a partial fraction

$$\frac{B}{(x-b)^m} \text{ will be } B\frac{(-1)^n m(m+1)\ldots(m+n-1)}{(x-b)^{m+n}}.$$

To find the nth d.c. of a partial fraction of the form (§ 40),

$$\frac{Px+Q}{(x-a)^2+\beta^2},$$

suppose it resolved into its conjugate imaginary partial fractions of the form,

$$\frac{A+iB}{x-a-i\beta}+\frac{A-iB}{x-a+i\beta}, \quad .$$

and then the nth d.c. will be

$$(-1)^n n! \left\{ \frac{A+iB}{(x-a-i\beta)^n}+\frac{A-iB}{(x-a+i\beta)^n} \right\}$$

$$=(-1)^n n! \frac{(A+iB)(x-a+i\beta)^n+(A-iB)(x-a-i\beta)^n}{\{(x-a)^2+\beta^2\}^n}.$$

This can be expressed in a real form by putting $x-a=\beta \cot \theta$, and then becomes, by De Moivre's theorem,

$$=(-1)^n n! \frac{(A+iB)(\cos n\theta+i\sin n\theta)+(A-iB)(\cos n\theta-i\sin n\theta)}{\{(x-a)^2+\beta^2\}^{\frac{1}{2}n}}$$

$$=2(-1)^n n! \frac{A\cos n\theta-B\sin n\theta}{\{(x-a)^2+\beta^2\}^{\frac{1}{2}n}}.$$

Also
$$\frac{d\log(x-a)}{dx}=\frac{1}{x-a},$$

and therefore
$$\frac{d^n\log(x-a)}{dx^n}=\frac{(-1)^{n-1}(n-1)!}{(x-a)^n}.$$

Similarly the nth d.c. of $\tan^{-1}x$ and $\tanh^{-1}x$ can be determined.

77. By § 4

$$\frac{d\sin x}{dx}=\cos x=\sin(x+\tfrac{1}{2}\pi),$$

therefore
$$\frac{d^2\sin x}{dx^2}=\sin(x+\tfrac{1}{2}2\pi),$$

and generally
$$\frac{d^n\sin x}{dx^n}=\sin(x+\tfrac{1}{2}n\pi).$$

Similarly $\dfrac{d^n \sin (px+q)}{dx^n} = p^n \sin (px+q+\tfrac{1}{2}n\pi),$

$$\dfrac{d^n \cos (px+q)}{dx^n} = p^n \cos (px+q+\tfrac{1}{2}n\pi).$$

Since $\dfrac{da^x}{dx} = a^x \log a$ (§ 26),

therefore generally $\dfrac{d^n a^x}{dx^n} = a^x (\log a)^n;$

and $\dfrac{d^n e^x}{dx^n} = e^x.$

Also (§ 32)

$$\dfrac{d^{2n} \sinh x}{dx^{2n}} = \sinh x, \quad \dfrac{d^{2n+1} \sinh x}{dx^{2n+1}} = \cosh x.$$

$$\dfrac{d^{2n} \cosh x}{dx^{2n}} = \cosh x, \quad \dfrac{d^{2n+1} \cosh x}{dx^{2n+1}} = \sinh x.$$

78. *Leibnitz's Theorem.*

We have already proved the rule for the differentiation of uv, the product of u and v, two given functions of x; namely (§ 34)

$$\dfrac{duv}{dx} = \dfrac{du}{dx}v + u\dfrac{dv}{dx}.$$

Differentiating again, each term on the right-hand side being a product gives rise to two terms; and

$$\dfrac{d^2uv}{dx^2} = \dfrac{d^2u}{dx^2}v + \dfrac{du}{dx}\dfrac{dv}{dx} + \dfrac{du}{dx}\dfrac{dv}{dx} + u\dfrac{d^2v}{dx^2}$$

$$= \dfrac{d^2u}{dx^2}v + 2\dfrac{du}{dx}\dfrac{dv}{dx} + u\dfrac{d^2v}{dx^2}.$$

Again $\dfrac{d^3uv}{dx^3} = \dfrac{d^3u}{dx^3}v + 3\dfrac{d^2u}{dx^2}\dfrac{dv}{dx} + 3\dfrac{du}{dx}\dfrac{d^2v}{dx^2} + u\dfrac{d^3v}{dx^3};$

and

$$\dfrac{d^4uv}{dx^4} = \dfrac{d^4u}{dx^4}v + 4\dfrac{d^3u}{dx^3}\dfrac{dv}{dx} + 6\dfrac{d^2u}{dx^2}\dfrac{d^2v}{dx^2} + 4\dfrac{du}{dx}\dfrac{d^3v}{dx^3} + u\dfrac{d^4v}{dx^4}.$$

We now perceive the law for any number of differentiations, by analogy with the Binomial Theorem; and the law can be proved by Mathematical Induction.

For assuming that

$$\frac{d^n uv}{dx^n} = \frac{d^n u}{dx^n}v + n\frac{d^{n-1}u}{dx^{n-1}}\frac{dv}{dx} + \frac{n(n-1)}{1.2}\frac{d^{n-2}u}{dx^{n-2}}\frac{d^2v}{dx^2} + \ldots\ldots$$

$$+ \frac{n(n-1)}{1.2}\frac{d^2u}{dx^2}\frac{d^{n-2}v}{dx^{n-2}} + n\frac{du}{dx}\frac{d^{n-1}v}{dx^{n-1}} + u\frac{d^n v}{dx^n}\ldots\ldots\ldots(1)$$

Differentiating again, each term on the right-hand side of (1) gives rise to two terms, of which the second of one term coalesces with the first of the next term; so that

$$\frac{d^{n+1}uv}{dx^{n+1}} = \frac{d^{n+1}u}{dx^{n+1}}v + (n+1)\frac{d^n u}{dx^n}\frac{dv}{dx} + \frac{(n+1)n}{1.2}\frac{d^{n-1}u}{dx^{n-1}}\frac{d^2v}{dx^2} + \ldots\ldots$$

$$+ \frac{(n+1)n}{1.2}\frac{d^2u}{dx^2}\frac{d^{n-1}v}{dx^{n-1}} + (n+1)\frac{du}{dx}\frac{d^n v}{dx^n} + u\frac{d^{n+1}v}{dx^{n+1}}\ldots(2)$$

If therefore the law expressed by (1) holds for n, it holds when n is changed into $n+1$, as expressed in (2).

But the law holds when n is 1, 2, 3, 4, and therefore it holds when n is 5, 6, ..., and generally any positive integer.

This law is called *Leibnitz's Theorem.*

Leibnitz's Theorem can be established by a *symbolical* proof, which is easily extended to the case of the differentiation of any number of factors; for if

$$y = uvw\ldots$$

when u, v, w,... are functions of x, then

$$\frac{dy}{dx} = \frac{du}{dx}vw\ldots + u\frac{dv}{dx}w\ldots + uv\frac{dw}{dx}\ldots + \ldots$$

$$= (D_1 + D_2 + D_3 + \ldots)uvw\ldots,$$

where D_1 represents the operation of differentiation on u, D_2 on v, D_3 on w....

Then, since these *operators* represented by D obey the same laws as algebraical quantities,

$$\frac{d^n y}{dx^n} = (D_1 + D_2 + D_3 + \ldots)^n uvw\ldots,$$

so that the coefficient of $\dfrac{d^p u}{dx^p}\dfrac{d^q v}{dx^q}\dfrac{d^r w}{dx^r}\ldots$ is equal to the coefficient of $x^p y^q z^r \ldots$ in the expansion of $(x+y+z+\ldots)^n$ by the Multinomial Theorem.

79. By Leibnitz's Theorem,

$$\frac{d^n e^{ax} y}{dx^n} = e^{ax} \left\{ a^n y + na^{n-1}\frac{dy}{dx} + \frac{n(n-1)}{1.2}a^{n-2}\frac{d^2 y}{dx^2} + \ldots \right\}$$

$$= e^{ax} \left\{ a^n + na^{n-1}\frac{d}{dx} + \frac{n(n-1)}{1.2}a^{n-2}\frac{d^2}{dx^2} + \ldots \right\} y$$

$$= e^{ax} \left(a + \frac{d}{dx} \right)^n y,$$

with the *symbolical* notation.

Therefore, if Fx denotes a rational function of x,

$$F\left(\frac{d}{dx}\right)e^{ax} y = e^{ax} F\left(a + \frac{d}{dx}\right)y,$$

a theorem of great use in *Differential Equations*.

Examples on Successive Differentiation.

1. $y = e^{x\cos a}\cos(x\sin a)$, $\dfrac{d^n y}{dx^n} = e^{x\cos a}\cos(x\sin a + na)$.

2. $y = e^{ax+b}\cos(px+q)$, $\dfrac{d^n y}{dx^n} = a^n(\sec a)^n e^{ax+b}\cos(px+q+na)$,

where $\tan a = \dfrac{p}{a}$.

3. $y = \cosh(ax+b)\cos(px+q)$,

$$\frac{d^n y}{dx^n} = a^n (\sec a)^n \{\cosh(ax+b)\cos(px+q)\cos na$$

$$- \sinh(ax+b)\sin(px+q)\sin na\},$$

if n is even; and find the value when n is odd.

4. $y = \dfrac{x}{(x-a)(x-b)(x-c)}$, find $\dfrac{d^n y}{dx^n}$.

5. $y^n \cos nx = a^n$, $\dfrac{d^2 y}{dx^2} + y = (n+1)\dfrac{y^{2n+1}}{a^{2n}}$.

6. $y = A\cos nx + B\sin nx$, $\dfrac{d^2 y}{dx^2} + n^2 y = 0$.

7. $y = A\cosh nx + B\sinh x$, or $ae^{nx} + be^{-nx}$,

$$\frac{d^2 y}{dx^2} - n^2 y = 0.$$

8. $y = x^{m-\frac{1}{2}}(A\cos nx + B\sin nx)$,

$$x^2 \frac{d^2 y}{dx^2} - (2m-1)x\frac{dy}{dx} + (n^2 x^2 + m^2 - \tfrac{1}{4})y = 0.$$

80. *Dynamical Applications.*

Suppose a body M, like a railway train, is moving in the straight line Ox (fig. 3), and that x denotes its distance from O at the time t; then (§ 13), if u denotes the velocity of the train, $u = \dfrac{dx}{dt}$.

The *acceleration* is defined to be the *rate of change of velocity per unit of time*, so that the acceleration of M is

$$\frac{du}{dt} = \frac{d^2 x}{dt^2}.$$

Also

$$\frac{du}{dt} = \frac{du}{dx}\frac{dx}{dt} = \frac{du}{dx}u = \tfrac{1}{2}\frac{du^2}{dx}.$$

Suppose, for example, the acceleration is constant and equal to f; then

$$\frac{du}{dt} = \frac{d^2x}{dt^2} = f.$$

Integrating with respect to t,

$$u = \frac{dx}{dt} = V + ft \dots\dots\dots\dots(1),$$

if V denotes the velocity when $t = 0$; so that V is the arbitrary constant to be added in integration (§ 37).

Integrating again with respect to t,

$$x = Vt + \tfrac{1}{2}ft^2, \dots\dots\dots\dots(2)$$

supposing $x = 0$, when $t = 0$.

Again,

$$\tfrac{1}{2}\frac{du^2}{dx} = f,$$

so that, integrating with respect to x,

$$\tfrac{1}{2}u^2 = \tfrac{1}{2}V^2 + fx \dots\dots\dots\dots(3);$$

and (1), (2), (3) are the equations of rectilinear motion with constant acceleration f.

For a retardation f is negative.

81. Suppose a body N to be projected vertically upwards from O with velocity V, and let y denote the height above O and v the upward velocity of N at the time t.

Then, if the resistance of the air is left out of account,

$$\frac{dv}{dt} = \frac{d^2y}{dt^2} = -g,$$

g denoting the acceleration of *gravity*, due to the attraction of the earth.

Therefore, integrating with respect to t,

$$v = \frac{dy}{dt} = V - gt,$$

and integrating again

$$y = Vt - \tfrac{1}{2}gt^2,$$

also

$$\tfrac{1}{2}v^2 = \tfrac{1}{2}V^2 - gy.$$

If the body N rises to the height h, and if T denotes the whole time of going up and coming down again, then $\frac{1}{2}T$ is the time of rising and also of falling, and $V = \frac{1}{2}gT$, $\frac{1}{2}V^2 = gh$.

Therefore $\qquad v = \frac{1}{2}gT - gt$,

and $\qquad\qquad y = \frac{1}{2}gt(T - t)$,

also $\qquad\qquad \frac{1}{2}v^2 = g(h - y)$.

82. *Vertical motion of a body when the resistance of the air is taken into account.*

Suppose a body like a sphere to be projected vertically in the air, and suppose the resistance of the air to vary as the nth power of the velocity; then the retardation due to the resistance of the air when the velocity of the body is v may be written $g\left(\dfrac{v}{w}\right)^n$, and then w is called the *terminal velocity* of the body, being the constant velocity with which the body would fall when the weight of the body is balanced by the resistance of the air, as observable in the terminal velocity of raindrops, hailstones, and meteorites.

When the body is moving *downwards* with velocity v, the equation of motion is

$$\frac{dv}{dt} = g - g\left(\frac{v}{w}\right)^n \dotfill (1)$$

and when the body is moving *upwards* with velocity v, the equation of motion is

$$\frac{dv}{dt} = -g - g\left(\frac{v}{w}\right)^n \dotfill (2)$$

Example 1.—Suppose $n=2$, and the time is reckoned from the instant the body is at the highest point.

Then for the *downward* motion of the body, equation (1) becomes

$$\frac{dv}{dt}=g\left(1-\frac{v^2}{w^2}\right),$$

and therefore

$$t=\frac{w}{g}\int_0^v \frac{d\frac{v}{w}}{1-\frac{v^2}{w^2}}=\frac{w}{g}\tanh^{-1}\frac{v}{w},$$

or

$$v=\frac{dy}{dt}=w\tanh\frac{gt}{w},$$

if y denotes the depth below the highest point at time t.

Therefore

$$y=w\int_0 \tanh\frac{gt}{w}dt=\frac{w^2}{g}\log\cosh\frac{gt}{w}.$$

For the *upward* motion of the body, equation (2) becomes

$$\frac{dv}{dt}=-g\left(1+\frac{v^2}{w^2}\right)$$

and therefore

$$t=-\frac{w}{g}\int_0^v \frac{d\frac{v}{w}}{1+\frac{v^2}{w^2}}=-\frac{w}{g}\tan^{-1}\frac{v}{w},$$

or

$$v=-\frac{dy}{dt}=-w\tan\frac{gt}{w}.$$

Therefore

$$y=w\int_0 \tan\frac{gt}{w}dt=\frac{w^2}{g}\log\sec\frac{gt}{w}.$$

The upward velocity is therefore infinite at an infinite depth below the highest point, but the whole time of ascent is finite, namely, $\frac{1}{2}\pi\frac{w}{g}$.

The downward velocity gradually increases from zero to w, the terminal velocity, which is reached in an infinite time and at an infinite depth below the highest point.

Example 2.—Suppose $n=4$; then equation (1) becomes

$$\frac{dv}{dt}=\tfrac{1}{2}\frac{dv^2}{dy}=g\left(1-\frac{v^4}{w^4}\right),$$

taking y as the independent variable.

Therefore

$$y=\tfrac{1}{2}\frac{w^2}{g}\int_0^{\frac{v^2}{w^2}}\frac{d\frac{v^2}{w^2}}{1-\frac{v^4}{w^4}}=\tfrac{1}{2}\frac{w^2}{g}\tanh^{-1}\frac{v^2}{w^2},$$

or

$$v^2=w^2\tanh\frac{2gy}{w^2},$$

$$v=\frac{dy}{dt}=w\sqrt{\left(\tanh\frac{2gy}{w^2}\right)}.$$

Therefore

$$t=\int_0\sqrt{\left(\coth\frac{2gy}{w^2}\right)}\frac{dy}{w}$$

$$=\tfrac{1}{2}\frac{w}{g}\left\{\tan^{-1}\sqrt{\left(\tanh\frac{2gy}{w^2}\right)}+\tanh^{-1}\sqrt{\left(\tanh\frac{2gy}{w^2}\right)}\right\}$$

(ex. 13, p. 126).

Equation (2) for the upward motion becomes

$$\frac{dv}{dt}=-\tfrac{1}{2}\frac{dv^2}{dy}=-g\left(1+\frac{v^4}{w^4}\right),$$

and therefore $v^2=w^2\tan\dfrac{2gy}{w^2}$,

$$v=-\frac{dy}{dt}=-w\sqrt{\left(\tan\frac{2gy}{w^2}\right)}.$$

Therefore

$$t = \tfrac{1}{2}\sqrt{2}\frac{w}{g}\left\{\sin^{-1}\sqrt{\left(\sin\frac{4gy}{w^2}\right)} + \sinh^{-1}\sqrt{\left(\sin\frac{4gy}{w^2}\right)}\right\}$$

(ex. 12, p. 126).

In the upward motion the velocity becomes infinite at a finite depth, $\dfrac{\pi w^2}{4g}$, below the highest point.

To complete the trajectory of the body, the velocity at this point must be supposed reversed, and the body projected downwards with infinite velocity; then at a depth y below the point of projection,

$$\frac{dv}{dt} = \tfrac{1}{2}\frac{dv^2}{dy} = g\left(1 - \frac{v^4}{w^4}\right),$$

$$v = \frac{dy}{dt} = w\sqrt{\left(\coth\frac{2gy}{w^2}\right)}.$$

83. *Equations of motion in a plane.*

If x, y are the co-ordinates at the time t of a point P moving in a plane (fig. 3), then (§ 13), $\dfrac{dx}{dt}$ and $\dfrac{dy}{dt}$ are the component velocities of P parallel to Ox and Oy.

Similarly, $\dfrac{d^2x}{dt^2}$ and $\dfrac{d^2y}{dt^2}$ are the component *accelerations* parallel to Ox and Oy, since the acceleration in any direction is defined to be the *rate of change of velocity in that direction per unit of time.*

84. Suppose, for instance, that a body P is projected from O (fig. 23) with velocity V at an angle a to the horizon, and suppose the resistance of the air to be left out of account.

Then, if the axis Ox is horizontal, and the axis Oy drawn vertically upwards, the equations of motion are

$$\frac{d^2x}{dt^2} = 0, \quad \frac{d^2y}{dt^2} = -g. \quad \cdot$$

Integrating with respect to t:

$$\frac{dx}{dt} = \text{a constant} = V \cos a,$$

$$\frac{dy}{dt} = \text{a constant} - gt = V \sin a - gt \,;$$

and integrating again

$$x = Vt \cos a,$$

$$y = Vt \sin a - \tfrac{1}{2}gt^2,$$

supposing $t = 0$ at O, the point of projection.

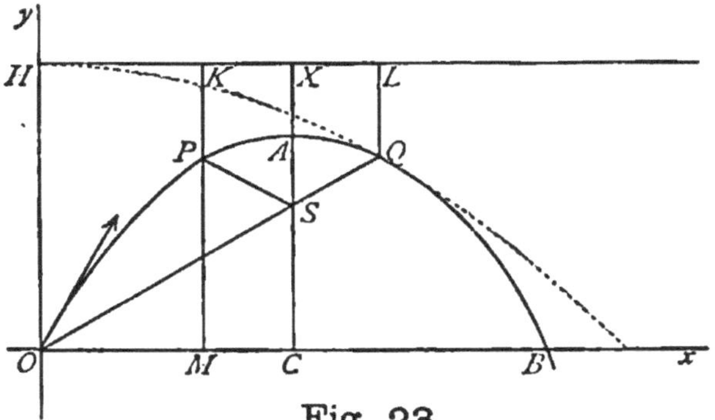

Fig. 23.

Therefore $t = \dfrac{x}{V \cos a}$, and substituting in the value of y,

$$y = x \tan a - \frac{gx^2}{2 V^2 \cos^2 a},$$

the equation of the trajectory; which will be found to be a parabola.

For writing it in the form

$$x^2 - \frac{2V^2x}{g}\sin a \cos a = -\frac{2V^2y}{g}\cos^2 a,$$

and completing the square in x;

$$\left(x - \frac{V^2}{g}\sin a \cos a\right)^2 = -\frac{2V^2}{g}\cos^2 a\left(y - \frac{V^2}{2g}\sin^2 a\right);$$

and comparing this with the equation

$$(x-h)^2 = -p(y-k),$$

which is the equation of a parabola as in fig. 23, of which the vertex is at (h, k), the latus rectum is p, and the concavity downwards; then

$$h = \frac{V^2}{g}\sin a \cos a = OC, \quad k = \frac{V^2}{2g}\sin^2 a = CA, \quad p = \frac{2V^2}{g}\cos^2 a.$$

If HK is the directrix of the parabola, then

$$OH = k + \tfrac{1}{4}p = \frac{V^2}{2g}.$$

Again

$$\frac{dx}{dt}\frac{d^2x}{dt^2} + \frac{dy}{dt}\frac{d^2y}{dt^2} = -g\frac{dy}{dt},$$

and integrating

$$\tfrac{1}{2}\left(\frac{dx^2}{dt^2} + \frac{dy^2}{dt^2}\right) = \text{a constant} - gy,$$

or

$$\tfrac{1}{2}v^2 = \tfrac{1}{2}V^2 - gy;$$

if v denotes the velocity at P.

The velocity v at any point P is therefore the velocity which would be acquired in falling freely from the level of the directrix; for

$$\tfrac{1}{2}v^2 = g \cdot OH - g \cdot MP = g \cdot PK.$$

To find the range OB, and the time of flight on the horizontal plane through O, put $y = 0$; then

$$0 = x \tan a - \frac{gx^2}{2V^2\cos^2 a},$$

$$0 = Vt \sin a - \tfrac{1}{2}gt^2.$$

Therefore, if the range is denoted by R,

$$R = \frac{2V^2}{g} \sin a \cos a = \frac{V^2}{g} \sin 2a,$$

and if the time of flight is denoted by T,

$$T = \frac{2V}{g} \sin a.$$

With a given velocity of projection V, the range R is a maximum when $\sin 2a = 1$, or $a = \tfrac{1}{4}\pi$.

Produce OS to meet the parabolic trajectory again in Q, and draw QL perpendicular to the directrix HK; then

$$OQ = OS + SQ = OH + QL,$$

so that the trajectory touches at Q a fixed parabola HQ, with focus at O and vertex at H; the parabola HQ is called the *envelope* of the trajectories like OPQ described with the same velocity of projection V, but varying elevation a.

The interior of the surface formed by the revolution of the parabola HQ round the vertical line OH will therefore be the whole space covered from the point O with the given velocity of projection V; and the section of this surface by any plane will be the area covered on that plane.

The maximum range on a line through O, like OQ, will therefore be obtained when OQ is a focal chord, and then the direction of projection bisects the angle HOQ.

For examples on the subject of parabolic motion, the reader is referred to treatises on Dynamics.

85. *Dynamical Equations with Polar Co-ordinates.*

Changing to polar co-ordinates r and θ, then (fig. 24)

$$x = r \cos \theta, \text{ and } y = r \sin \theta.$$

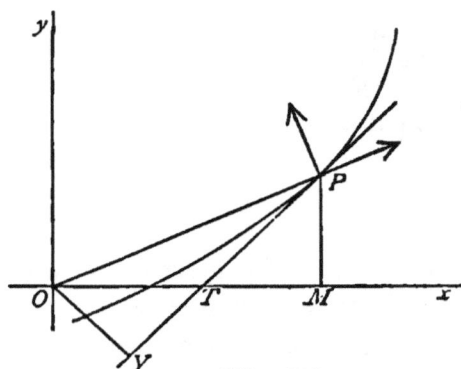

Fig. 24.

Differentiating with respect to t,

$$\frac{dx}{dt} = \frac{dr}{dt} \cos \theta - r \sin \theta \frac{d\theta}{dt},$$

$$\frac{dy}{dt} = \frac{dr}{dt} \sin \theta + r \cos \theta \frac{d\theta}{dt}.$$

Therefore the component velocity in the direction OP, called the *radial velocity,*

$$\frac{dx}{dt} \cos \theta + \frac{dy}{dt} \sin \theta = \frac{dr}{dt} ;$$

and the component velocity perpendicular to OP, called the *transversal velocity,*

$$-\frac{dx}{dt} \sin \theta + \frac{dy}{dt} \cos \theta = r \frac{d\theta}{dt},$$

as before (§ 18).

Differentiating again,

$$\frac{d^2x}{dt^2} = \left(\frac{d^2r}{dt^2} - r\frac{d\theta^2}{dt^2}\right)\cos\theta - \left(2\frac{dr}{dt}\frac{d\theta}{dt} + r\frac{d^2\theta}{dt^2}\right)\sin\theta,$$

$$\frac{d^2y}{dt^2} = \left(\frac{d^2r}{dt^2} - r\frac{d\theta^2}{dt^2}\right)\sin\theta + \left(2\frac{dr}{dt}\frac{d\theta}{dt} + r\frac{d^2\theta}{dt^2}\right)\cos\theta.$$

Therefore the component *radial acceleration*

$$\frac{d^2x}{dt^2}\cos\theta + \frac{d^2y}{dt^2}\sin\theta = \frac{d^2r}{dt^2} - r\frac{d\theta^2}{dt^2},$$

and the component *transversal acceleration*

$$-\frac{d^2x}{dt^2}\sin\theta + \frac{d^2y}{dt^2}\cos\theta = 2\frac{dr}{dt}\frac{d\theta}{dt} + r\frac{d^2\theta}{dt^2},$$

which is usually written in the form $\frac{1}{r}\frac{d}{dt}\left(r^2\frac{d\theta}{dt}\right)$, to which it is equivalent, as may be seen by differentiating this last expression.

Suppose, for instance, that P describes a circle round O as centre; then r is constant and $\frac{dr}{dt} = 0$; so that the radial acceleration is $-r\frac{d\theta^2}{dt^2}$ and the transversal or tangential acceleration is $r\frac{d^2\theta}{dt^2}$.

Also, if v denotes the velocity in the circle, $v = \frac{ds}{dt} = r\frac{d\theta}{dt}$, so that the tangential acceleration is $\frac{dv}{dt}$ or $\frac{d^2s}{dt^2}$ or $v\frac{dv}{ds}$, the same as for rectilinear motion; and the central or normal acceleration in the direction PO is $r\frac{d\theta^2}{dt^2} = v\frac{d\theta}{dt} = \frac{v^2}{r}$.

To realize this circular motion practically, suppose a particle, suspended from a fixed point by a fine string, to be projected with velocity v so as to describe a horizontal circle of radius r under gravity, neglecting the resistance of the air.

K

Denoting by a the constant inclination of the string to the vertical, and by Q the tension of the string, then

$$\frac{v^2}{r} : g :: Q \sin a : Q \cos a,$$

or $$v^2 = gr \tan a.$$

Now if $2T$ denotes the time of revolution in the circle,

$$v = \frac{\pi r}{T};$$

therefore $$v^2 = \frac{\pi^2}{T^2} r^2 = gr \tan a,$$

or $$T = \pi \sqrt{\frac{r \cot a}{g}} = \pi \sqrt{\frac{l \cos a}{g}},$$

l denoting the length of the string; and T therefore approaches the limit $\pi \sqrt{\frac{l}{g}}$, as a becomes indefinitely diminished.

86. Suppose a particle P moves in a *field of force* in which the radial and transversal accelerations due to the field of force are represented by R and T respectively; then the equations of motion are

$$\frac{d^2r}{dt^2} - r\frac{d\theta^2}{dt^2} = R \dots\dots\dots\dots (1)$$

$$\frac{1}{r}\frac{d}{dt}\left(r^2\frac{d\theta}{dt}\right) = T \dots\dots\dots\dots (2)$$

Now $r^2 \frac{d\theta}{dt}$ is generally denoted by h, and then $h = 2\frac{dA}{dt}$ (§ 60), twice the rate the area is described by the vector OP about O.

Change the independent variable in the equations of motion from t to θ, and denote r by $\frac{1}{u}$.

Then since $r^2 \dfrac{d\theta}{dt} = h,$

therefore $\dfrac{d\theta}{dt} = hu^2 ;$

and equation (2) becomes

$$\dfrac{dh}{dt} = \dfrac{T}{u}, \text{ or } \dfrac{dh}{d\theta} = \dfrac{T}{hu^3}, \text{ or } \tfrac{1}{2}\dfrac{dh^2}{d\theta} = \dfrac{T}{u^3}.$$

Again $\dfrac{dr}{dt} = \dfrac{dr}{d\theta}\dfrac{d\theta}{dt} = -\dfrac{1}{u^2}\dfrac{du}{d\theta}hu^2 = -\dfrac{du}{d\theta}h ;$

so that $\dfrac{d^2r}{dt^2} = -\dfrac{d^2u}{d\theta^2}\dfrac{d\theta}{dt}h - \dfrac{du}{d\theta}\dfrac{dh}{dt}$

$$= -\dfrac{d^2u}{d\theta^2}h^2u^2 - \dfrac{du}{d\theta}\dfrac{T}{u},$$

and $r\dfrac{d\theta^2}{dt^2} = h^2u^3.$

Therefore equation (1) becomes

$$-\dfrac{d^2u}{d\theta^2}h^2u^2 - \dfrac{du}{d\theta}\dfrac{T}{u} - h^2u^3 = R,$$

or $\dfrac{d^2u}{d\theta^2} + u = -\dfrac{R}{h^2u^2} - \dfrac{T}{h^2u^3}\dfrac{du}{d\theta},$

the *differential equation of the orbit* of *P*.

87. *Central Orbits.*

For a central field of force in which the attraction is always directed to the origin O, $T = 0$, and h is therefore constant.

Denoting the central acceleration due to the attraction towards O by P, so that $P = -R$, then

$$\dfrac{d^2u}{d\theta^2} + u = \dfrac{P}{h^2u^2},$$

or $P = h^2u^2\left(\dfrac{d^2u}{d\theta^2} + u\right),$

whence the required value of P is found when the equation of the orbit is given.

Examples.—(1) The polar equation of a conic section with a focus at the origin being

$$\frac{l}{r} \text{ or } lu = 1 + e \cos \theta,$$

then

$$l\left(\frac{d^2u}{d\theta^2} + u\right) = 1,$$

and therefore $\quad P = \dfrac{h^2}{l}u^2$, which varies as u^2 or $\dfrac{1}{r^2}$.

Thus Newton's Law of Gravitation by which the Sun attracts the planets with intensity inversely proportional to the square of the distance is deduced from Kepler's Law, that the planets describe ellipses of which the Sun occupies a focus.

(2) Prove that, for a conic section with the centre at the origin, P varies as r.

(3) Prove that P varies as u^3 in the curves (*Cotes's Spirals*) (i.) $au = \cosh n\theta$, (ii.) $au = \exp n\theta$ (the equiangular spiral, p. 98), (iii.) $au = \sinh n\theta$, (iv.) $au = n\theta$ (the hyperbolic spiral, p. 98), (v.) $au = \sin n\theta$.

(4) Prove that P varies as r^{-2n-3} in $r^n = a^n \cos n\theta$.

(5) Prove that

(i.) P varies as u^5 in $au = \tanh \dfrac{\theta}{\sqrt{2}}$ or $\coth \dfrac{\theta}{\sqrt{2}}$;

(ii.) P varies as u^4 in $au = \dfrac{\cosh\theta - 2}{\cosh\theta + 1}$ or $\dfrac{\cosh\theta + 2}{\cosh\theta - 1}$;

(iii.) P varies as u^7 in $a^2u^2 = \dfrac{\cosh 2\theta - 1}{\cosh 2\theta + 2}$ or $\dfrac{\cosh 2\theta + 1}{\cosh 2\theta - 2}$.

In the curves of this last example, $au = 1$ or $r = a$ when $\theta = \infty$; so that after an infinite number of revolutions round the origin the curves approach to coincidence with the circle $r = a$, which is therefore called an *asymptotic circle*.

A particle describing this circle ' freely under the central attractions of (i.), (ii.), or (iii.) would be unstable, and be found ultimately describing one of the corresponding curves.

The inverse problem to determine the orbit when the central attraction is given is more complicated, but the general method pursued is as follows.

Supposing P a given function of r or u, then

$$\frac{d^2u}{d\theta^2} + u = \frac{P}{h^2u^2}.$$

Multiplying by $\frac{du}{d\theta}$,

$$\frac{du}{d\theta}\frac{d^2u}{d\theta^2} + u\frac{du}{d\theta} = \frac{P}{h^2u^2}\frac{du}{d\theta},$$

and integrating with respect to θ,

$$\tfrac{1}{2}\frac{du^2}{d\theta^2} + \tfrac{1}{2}u^2 = \int \frac{P}{h^2u^2}du + C,$$

or

$$\frac{du^2}{d\theta^2} = 2\int \frac{P}{h^2u^2}du - u^2 + 2C;$$

so that

$$\theta = \int \left(2\int \frac{P}{h^2u^2}du - u^2 + 2C\right)^{-\frac{1}{2}}du,$$

the equation of the orbit, connecting θ and u.

For examples the reader is referred to the treatises on analytical Dynamics by Tait and Steele, Besant, or Williamson.

88. *Geometrical Illustrations of Successive Differen-*
tiation.

Curvature. The *curvature* at any point of a plane
curve is defined to be the rate at which the curve is
bending, or the rate at which the tangent is revolving
per unit length of the curve.

The angle (in circular measure) between the tangents
at the ends of a finite arc of a plane curve is called the
whole curvature of the arc, and the ratio of the whole
curvature to the length of the arc is called the *average*
curvature of the arc.

Suppose that, in going from a point P to an adjacent
point P' on an arc of length Δs, the tangent (or normal)
has turned through the angle $\Delta \psi$, then $\Delta \psi$ is the whole
curvature, and $\dfrac{\Delta \psi}{\Delta s}$ is the average curvature of the arc
PP'; and making P' move up on the curve to coincidence
with P, the curvature at P is $\operatorname{lt} \dfrac{\Delta \psi}{\Delta s} = \dfrac{d\psi}{ds}$.

To measure curvature a circle is employed, because the
curvature of a circle is the same at every point, and by
varying the radius the circle may be made to have any
required curvature.

For if PP' is the arc of a circle, then $\dfrac{\Delta s}{\Delta \psi}$ is always
equal to the radius (§ 5), and therefore the curvature of a
circle, measured by $\dfrac{\Delta \psi}{\Delta s}$, is the reciprocal of the radius.

The *circle of curvature* or the *osculating circle* at any
point of a curve is defined to be the circle which touches
and has the same curvature as the curve at the point.

89. Any small arc PP' may be considered as ultimately coincident with the arc of the circle of curvature, and therefore Q, the point of intersection of the normal at P' with the normal at P (fig. 25), is ultimately the *centre of curvature* at P, and then QP is called the *radius of curvature*, and denoting it by ρ, then $\rho = \mathrm{lt}\,\dfrac{\Delta s}{\Delta \psi} = \dfrac{ds}{d\psi}$.

The curvature at any point is therefore $\dfrac{1}{\rho}$ or $\dfrac{d\psi}{ds}$.

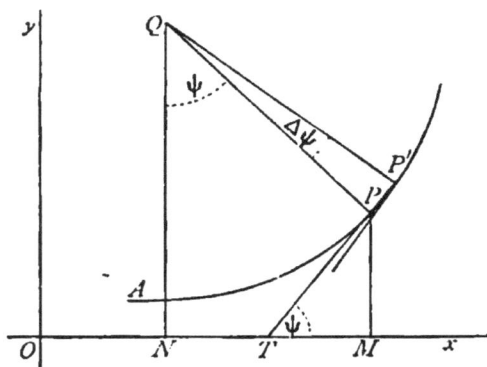

Fig. 25.

Now $\qquad \tan \psi = \dfrac{dy}{dx}$, or $\psi = \tan^{-1}\dfrac{dy}{dx}$;

therefore $\qquad \dfrac{d\psi}{dx} = \dfrac{\dfrac{d^2y}{dx^2}}{1 + \dfrac{dy^2}{dx^2}}$,

also $\qquad \dfrac{ds}{dx} = \sqrt{\left(1 + \dfrac{dy^2}{dx^2}\right)}$, (§ 13);

and
$$\rho = \frac{ds}{d\psi}$$

$$= \frac{\dfrac{ds}{dx}}{\dfrac{d\psi}{dx}} = \frac{\left(1 + \dfrac{dy^2}{dx^2}\right)^{\frac{3}{2}}}{\dfrac{d^2y}{dx^2}} \; ,$$

the expression for ρ in terms of y, $\dfrac{dy}{dx}$ and $\dfrac{d^2y}{dx^2}$.

90. Taking t as the independent variable, then

$$\psi = \tan^{-1}\frac{dy}{dx} = \tan^{-1}\frac{\dfrac{dy}{dt}}{\dfrac{dx}{dt}} \quad (\S\ 14),$$

and
$$\frac{d\psi}{dt} = -\frac{\dfrac{dx}{dt}\dfrac{d^2y}{dt^2} - \dfrac{d^2x}{dt^2}\dfrac{dy}{dt}}{\dfrac{dx^2}{dt^2} + \dfrac{dy^2}{dt^2}} \; ;$$

also ($\S$ 13)
$$\frac{ds}{dt} = \sqrt{\left(\frac{dx^2}{dt^2} + \frac{dy^2}{dt^2}\right)},$$

therefore
$$\rho = \frac{\left(\dfrac{dx^2}{dt^2} + \dfrac{dy^2}{dt^2}\right)^{\frac{3}{2}}}{\dfrac{dx}{dt}\dfrac{d^2y}{dt^2} - \dfrac{d^2x}{dt^2}\dfrac{dy}{dt}}.$$

We are now said to have *changed the independent variable from x to t* in the expression for ρ.

Since
$$v^2 = \frac{dx^2}{dt^2} + \frac{dy^2}{dt^2},$$

and
$$\frac{dx}{dt} = v \cos \psi, \quad \frac{dy}{dt} = v \sin \psi,$$

therefore
$$\rho = \frac{v^2}{\cos \psi \frac{d^2y}{dt^2} - \sin \psi \frac{d^2x}{dt^2}},$$

or $\dfrac{v^2}{\rho} = \cos \psi \dfrac{d^2y}{dt^2} - \sin \psi \dfrac{d^2x}{dt^2}$, the *normal component* of the acceleration of P.

The tangential component of the acceleration
$$\cos \psi \frac{d^2x}{dt^2} + \sin \psi \frac{d^2y}{dt^2}$$
$$= \frac{1}{v}\left(\frac{dx}{dt}\frac{d^2x}{dt^2} + \frac{dy}{dt}\frac{d^2y}{dt^2}\right) = \frac{dv}{dt} = v\frac{dv}{ds} = \tfrac{1}{2}\frac{dv^2}{ds},$$

the same as for rectilinear motion; and $\dfrac{v^2}{\rho}$, the normal acceleration, is the same as for motion in the circle of curvature (§ 85).

91. Taking s, the arc of the curve, for the independent variable, then since (§ 12)
$$\frac{dx}{ds} = \cos \psi, \quad \frac{dy}{ds} = \sin \psi;$$

therefore, differentiating with respect to s,
$$\frac{d^2x}{ds^2} = -\sin \psi \frac{d\psi}{ds} = -\frac{dy}{ds}\frac{1}{\rho},$$
$$\frac{d^2y}{ds^2} = \cos \psi \frac{d\psi}{ds} = \frac{dx}{ds}\frac{1}{\rho}.$$

Squaring and adding
$$\left(\frac{d^2x}{ds^2}\right)^2 + \left(\frac{d^2y}{ds^2}\right)^2 = \frac{1}{\rho^2};$$

also
$$\rho = -\frac{\dfrac{dy}{ds}}{\dfrac{d^2x}{ds^2}} = \frac{\dfrac{dx}{ds}}{\dfrac{d^2y}{ds^2}};$$

and
$$\cos \psi = \rho \frac{d^2y}{ds^2}, \quad \sin \psi = -\rho \frac{d^2x}{ds^2}.$$

If a, β are the co-ordinates of Q, the *centre of curvature* at P,

$$a = x - \rho \sin \psi = x + \rho^2 \frac{d^2x}{ds^2},$$

$$\beta = y + \rho \cos \psi = y + \rho^2 \frac{d^2y}{ds^2}.$$

With x for independent variable,

$$x - a = \rho \sin \psi = \frac{ds}{d\psi} \frac{dy}{ds} = \frac{\frac{dy}{dx}}{\frac{d\psi}{dx}} = \frac{\frac{dy}{dx}\left(1 + \frac{dy^2}{dx^2}\right)}{\frac{d^2y}{dx^2}},$$

$$y - \beta = -\rho \cos \psi = -\frac{ds}{d\psi} \frac{dx}{ds} = -\frac{1}{\frac{d\psi}{dx}} = -\frac{1 + \frac{dy^2}{dx^2}}{\frac{d^2y}{dx^2}}.$$

92. *The Evolute and Involute.*

Since $\qquad a = x - \rho \sin\psi, \quad \beta = y + \rho \cos \psi$;

then differentiating with respect to s,

$$\frac{da}{ds} = \frac{dx}{ds} - \frac{d\rho}{ds} \sin \psi - \rho \cos \psi \frac{d\psi}{ds} = -\frac{d\rho}{ds} \sin \psi,$$

$$\frac{d\beta}{ds} = \frac{dy}{ds} + \frac{d\rho}{ds} \cos \psi - \rho \sin \psi \frac{d\psi}{ds} = \frac{d\rho}{ds} \cos \psi \; ;$$

and therefore

$$\frac{d\beta}{da} = -\cot \psi = \tan \left(\tfrac{1}{2}\pi + \psi\right).$$

The locus of (a, β) the centre of curvature Q, is called the *evolute* of the curve AP; and the preceding equation shows that the tangent to the evolute at Q is QP, the normal to the curve at P.

Also $\qquad \dfrac{da^2}{ds^2} + \dfrac{d\beta^2}{ds^2} = \dfrac{d\rho^2}{ds^2},$ or $\dfrac{d\sigma^2}{ds^2} = \dfrac{d\rho^2}{ds^2},$

if σ denotes the length of the arc of the evolute, measured from a fixed point.

Therefore $\quad \dfrac{d\sigma}{ds} = \pm \dfrac{d\rho}{ds}$, and $\sigma =$ a constant $\pm \rho$.

(i.) Suppose $\sigma = \rho - l$; then the curve AP can be described by the end of a string unwrapped from the corresponding arc BQ of the evolute (fig. 26, i.).

(ii.) Suppose $\sigma = l - \rho$; then the curve AP can be described by the end of a string which is wrapped on the corresponding arc BQ of the evolute (fig. 26, ii.).

In each case, σ denotes the length of the arc BQ of the evolute, and $l = AB$, the radius of curvature at A, and $\rho = PQ$, the radius of curvature at P.

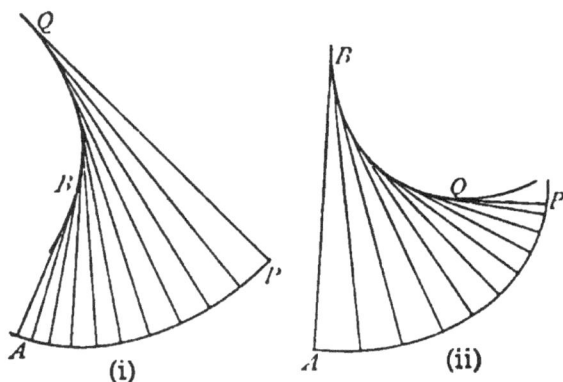

Fig. 26.

Relatively to the curve BQ, the curve AP is called an *involute;* and by varying the length of the string, any number of involutes can be obtained, all *parallel* curves; but a curve, like AP, has only one evolute, like BQ.

Thus the *tractrix* is an involute of the *catenary* (exs. 91, 92, p. 58).

93. *The cycloid and its evolute, and cycloidal vibrations.*

The *cycloid* is a curve *OP* (fig. 27) described by a point *P* fixed to the circumference of a circle which rolls on a straight line.

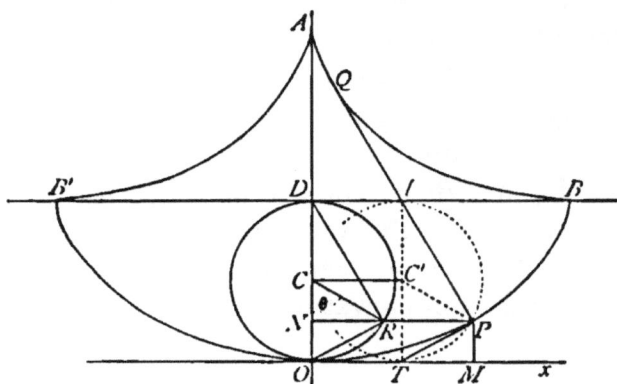

Fig.27

When the centre of the circle has advanced by rolling on *DB* from *C* to *C'*, the circle will have turned through an angle θ, where $CC'=a\theta$, a denoting the radius of the circle; so that if O, the point from which P starts when at the end of the diameter *DC* is taken as origin, and Ox drawn parallel to *DB*, then if x, y denote the co-ordinates of P,

$$x = a(\theta+\sin\theta),\ y = a \operatorname{vers}\theta\,;$$

so that
$$x = a \operatorname{vers}^{-1}\frac{y}{a} + \sqrt{(2ay-y^2)} \dots\dots\dots\dots(1)$$

The rotation and translation of the circle may be considered independently, by supposing the circle first to turn through an angle θ, carrying O to R, and then to slide on *DB* a distance $a\theta$, carrying R to P; thus the cycloid is constructed by producing the ordinate NR of the circle to P, and making $RP = \text{arc } OR$.

The curve described by P' when NP' is made equal to the arc OR is called the *companion of the cycloid*, but then

$$x = a\theta, \; y = a \text{ vers } \theta ;$$

or

$$y = a \text{ vers } \frac{x}{a},$$

a curve of the same form as (§ 16)

$$y = a \cos \frac{x}{a} \text{ or } y = a \sin \frac{x}{a}.$$

Differentiating in the cycloid with respect to θ,

$$\frac{dx}{d\theta} = a(1 + \cos \theta), \; \frac{dy}{d\theta} = a \sin \theta ;$$

so that

$$\frac{dy}{dx} = \frac{\sin \theta}{1 + \cos \theta} = \tan \tfrac{1}{2}\theta,$$

and therefore the angle xTP or $\psi = \tfrac{1}{2}\theta$; which proves that TP is parallel and therefore equal to OR.

This can be seen immediately from the fact that the rolling circle in any position is turning instantaneously about I, the point of contact with DB, and therefore IP is the normal and TP the tangent to the cycloid at P, where IT is the diameter of the rolling circle.

Again

$$\frac{ds^2}{d\theta^2} = \frac{dx^2}{d\theta^2} + \frac{dy^2}{d\theta^2}$$

$$= a^2\{(1 + \cos \theta)^2 + \sin^2\theta\} = 4a^2\cos^2\tfrac{1}{2}\theta,$$

so that

$$\frac{ds}{d\theta} = 2a \cos \tfrac{1}{2}\theta,$$

and therefore $\quad s = 4a \sin \tfrac{1}{2}\theta = 4a \sin \psi \dots\dots\dots\dots(2)$

s denoting the length of the arc OP, which is therefore equal to $2TP$.

The value of ρ, the radius of curvature at P, can be obtained from the formula (§ 90)

$$\rho = \frac{\left(\dfrac{dx^2}{d\theta^2} + \dfrac{dy^2}{d\theta^2}\right)^{\frac{3}{2}}}{\dfrac{dx}{d\theta}\dfrac{d^2y}{d\theta^2} - \dfrac{d^2x}{d\theta^2}\dfrac{dy}{d\theta}},$$

and will be found to be $4a\cos\frac{1}{2}\theta$; but this can be obtained immediately from equation (2), for

$$\rho = \frac{ds}{d\psi} = 4a\cos\psi = 4a\cos\tfrac{1}{2}\theta.$$

Therefore $PQ = 2PI$, if Q is the centre of curvature at P, and this shows that the evolute AQ is an equal cycloid.

For a, β denoting the co-ordinates of Q, then (§ 91)

$$a = x - \rho\sin\psi = a(\theta + \sin\theta) - 4a\cos\tfrac{1}{2}\theta\sin\tfrac{1}{2}\theta$$
$$= a(\theta - \sin\theta),$$
$$\beta = y + \rho\cos\psi = a(1 - \cos\theta) + 4a\cos^2\tfrac{1}{2}\theta$$
$$= 3a + a\cos\theta;$$

which proves that, as P describes the cycloid OPB, Q describes an equal cycloid AQB, produced by rolling a circle of radius a on the straight line through A parallel to DB.

This, the first problem of an evolute, was invented by Huyghens, with the object of making a particle vibrate in a cycloid; for if the cycloid BOB' is cut out of some material and divided at O, and $B'O$ is placed in the position AB, and OB in the position $B'A$, in a vertical plane with BB' horizontal, a particle at O hanging vertically from A by a fine string of length OA will, if made to vibrate in the vertical plane of the figure, describe an arc of the cycloid BOB'.

94. The peculiarity of these vibrations under gravity is that the time of vibration is the same for all arcs of vibration; this property is called the *isochronism of the cycloid.*

For resolving tangentially at P,

$$\tfrac{1}{2}\frac{dv^2}{ds} = \text{resolved part of gravity along the tangent}$$

$$= -g \sin \psi = -g\frac{s}{l},$$

if $l = 4a$, the length of the string.

Integrating with respect to s,

$$\tfrac{1}{2}v^2 = \tfrac{1}{2}\frac{g}{l}(s_1{}^2 - s^2),$$

supposing P to be drawn aside from O through an arc s_1, and then let go.

Therefore $\quad v = \dfrac{ds}{dt} = -\sqrt{\dfrac{g}{l}}\sqrt{(s_1{}^2 - s^2)},$

the negative sign being taken because P begins moving towards O; and

$$t - \tau = \sqrt{\frac{l}{g}}\int_s^{s_1} \frac{ds}{\sqrt{(s_1{}^2 - s^2)}} = \sqrt{\frac{l}{g}}\cos^{-1}\frac{s}{s_1},$$

or $\qquad\qquad s = s_1\cos\sqrt{\dfrac{g}{l}}(t - \tau),$

supposing τ to denote the instant of time when P is let go.

If T denotes the time of vibration from rest to rest,

then $\sqrt{\dfrac{g}{l}}(t - \tau)$ increases by π, while $t - \tau$ increases by T,

so that $\qquad\qquad s = s_1\cos\dfrac{\pi}{T}(t - \tau),$

and $\qquad\qquad T = \pi\sqrt{\dfrac{l}{g}},$

which is independent of s_1, and therefore the same for all arcs of vibration; which proves the isochronism of the cycloid.

When s_1, the *amplitude* of vibration, is small, the arc of vibration in the cycloid may be considered coincident with the arc of the circle of curvature at O, a circle of radius l and centre A; so that the time of a small plane vibration of a simple pendulum of length l is thus proved to be $\pi\sqrt{\dfrac{l}{g}}$, the same as the time of a semi-revolution in a small horizontal circle (§ 85).

95. *Harmonic Vibrations.*

If a point P describes the circle (fig. 12) with constant velocity in the periodic time $2T$, then

$$\frac{\theta}{\pi} = \frac{t-\tau}{T},$$

supposing P is at A at the instant of time denoted by τ; and then

$$OM = x = a \cos \theta = a \cos \frac{\pi}{T}(t-\tau).$$

The point M vibrates between A and A' in the time T, and M is said to perform a *harmonic vibration*, and a is called the *amplitude*, T the *time of vibration*, and τ is called the *epoch*, being one of the instants at which M is at A.

Thus, in the preceding article, P makes a harmonic vibration on the cycloid.

If

$$x = a \cos \frac{\pi}{T}(t-\tau),$$

then

$$\frac{d^2x}{dt^2} = -\frac{\pi^2}{T^2}x,$$

so that the point M vibrates as if *attracted* to O with intensity proportional to the distance from O.

The small vibrations of elastic bodies producing musical notes are of this nature, whence these vibrations are called *harmonic.*

Again, if θ denotes the excentric angle of a point P on the ellipse (fig. 13), then

$$x = a \cos \frac{\pi}{T}(t - \tau), \quad y = b \sin \frac{\pi}{T}(t - \tau);$$

so that the motion of P on the ellipse may be supposed compounded of two harmonic vibrations in directions parallel to Ox and $Oy.$

Also $\qquad \dfrac{d^2x}{dt^2} = -\dfrac{\pi^2}{T^2}x, \ \dfrac{d^2y}{dt^2} = -\dfrac{\pi^2}{T^2}y;$

so that P then moves on the ellipse as if attracted to O with intensity proportional to the distance from $O.$

This case can be exhibited practically by Lissajous' method, as explained in Ganot's Physics (§ 278), by noticing the curve formed on a screen by a spot of light reflected from mirrors on two tuning forks in unison, vibrating in planes perpendicular to each other.

Here, in the most general case,

$$x = a \cos \frac{\pi}{T}(t - \tau), \quad y = b \sin \frac{\pi}{T}(t - \tau'),$$

with different epochs τ and τ'; and eliminating $t,$

$$\frac{xy}{ab} - \sqrt{\left(1 - \frac{x^2}{a^2}\right)}\sqrt{\left(1 - \frac{y^2}{b^2}\right)} = \sin \frac{\pi}{T}(\tau - \tau'),$$

and rationalizing,

$$\frac{x^2}{a^2} - 2\frac{xy}{ab}\sin\frac{\pi}{T}(\tau - \tau') + \frac{y^2}{b^2} = \cos^2\frac{\pi}{T}(\tau - \tau'),$$

the equation of an ellipse; and by gradually varying $\tau - \tau'$, the difference of epochs, by putting the tuning forks slightly out of tune, the different figures are obtained in succession.

L

In the analogous problem for the hyperbola (fig. 14),

$$x = a \cosh u, \ y = a \sinh u,$$

where

$$u = m(t - \tau) ;$$

and then

$$\frac{d^2x}{dt^2} = m^2x, \ \frac{d^2y}{dt^2} = m^2y ;$$

so that P moves on the hyperbola as if *repelled* from O with intensity proportional to the distance from O.

96. *Intrinsic Equation of a Curve.*

Equation (2) (§ 93) connecting s and ψ is called the *intrinsic equation* of the cycloid.

Generally the relation $s = f\psi$ for any curve, connecting s the length of any arc measured from a fixed point, and ψ the whole curvature of the arc (§ 88), is called the *intrinsic equation* of the curve, and was invented by Whewell; the equation is called *intrinsic* because it is independent of a system of co-ordinates.

Thus $s = a\psi$ is the intrinsic equation of a circle of radius a; and $s = l \sin \psi$ is the intrinsic equation of a cycloid, generated by the rolling of a circle of radius $\frac{1}{4}l$.

Supposing ψ continually to increase, the complete cycloid will be composed of a number of equal branches coming to a point on the straight line DB in what are called *cusps* (fig. 28, i.).

More generally the curve whose intrinsic equation is

$$s = l \sin m\psi$$

will consist of a number of equal branches with equidistant cusps now arranged on a circle.

If $m < 1$, the curve lies outside the circle of cusps, and is an *epicycloid*—that is, the curve described by a point on the circumference of a circle rolling outside a fixed circle (fig. 28, ii.).

If $m > 1$, the curve lies inside the circle of cusps, and is *a hypocycloid*—that is, the curve described by a point on the circumference of a circle rolling inside a fixed circle (fig. 28, iii.).

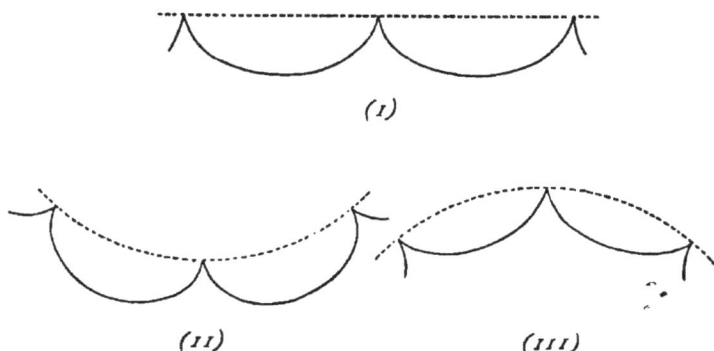

(I)

(II)   *(III)*

Fig.28

If $s = f\psi$ is the intrinsic equation of a curve, then

$$\sigma = \frac{ds}{d\psi} = f'\psi$$

is the intrinsic equation of the evolute (§ 92), measuring σ from the point where $\frac{ds}{d\psi} = 0$; and conversely

$$s = \int f\psi\, d\psi + C$$

is the intrinsic equation of an involute.

For instance, if $\quad s = l \sin m\psi$,

then $\qquad\qquad \sigma = \frac{ds}{d\psi} = ml \cos m\psi$,

which proves that the evolute of an epi- or hypocycloid is a similar epi- or hypo-cycloid.

Also $s = \frac{1}{2}a\psi^2$ is the intrinsic equation of the involute of a circle.

Examples.—Prove that

(1) $s = \frac{2}{3}a(\sec^3\psi - 1)$ is the intrinsic equation of the semi-cubical parabola (ex. 1, p. 95);

(2) $s = \frac{3}{4}a \sin 2\psi$ of $x^{\frac{2}{3}} + y^{\frac{2}{3}} = a^{\frac{2}{3}}$ (ex. 3);

(3) $s = a \tan \psi$, of the catenary (ex. 4);

(4) $s = a \log \sec \psi$, of the tractrix (ex. 5);

(5) $\psi = \gd \dfrac{s}{a}$, of the curve $y = a \log \sec \dfrac{x}{a}$ (ex. 6);

(6) $s = 4a \sin \frac{1}{3}\psi$, of the cardioid (ex. 8).

(7) Draw the curves (i.) $s = l \sin \frac{1}{2}\psi$, (ii.) $\psi = m \sin \dfrac{s}{a}$,

(iii.) $\psi = m \sin \dfrac{s^2}{a^2}$, (iv.) $\psi = \dfrac{s}{a} + m \sin \dfrac{s}{b}$.

(Whewell, *Cambridge Phil. Trans.*, vol. 8.)

97. *Circle of Curvature.*—Another way of obtaining the circle of curvature at a point P of a curve is to

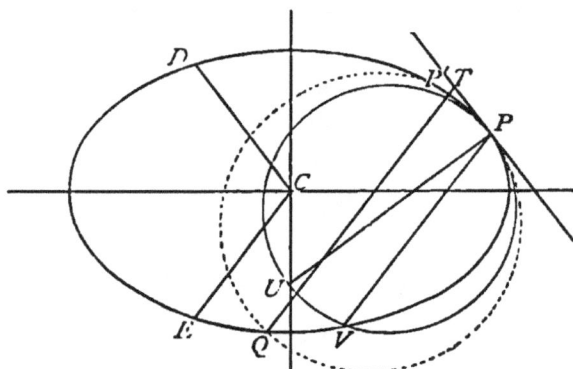

Fig. 29.

regard it as the ultimate position of the circle which touches the curve at P and cuts the curve again at a consecutive point P', when P' is moved up to coincidence with P.

For example, suppose the curve is a conic section, say an ellipse (fig. 29), and let the circle cut the ellipse again at Q.

Produce QP' to meet the tangent at P in T; then from the property of the circle

$$TP^2 = TP' \cdot TQ;$$

and from the property of the ellipse

$$\frac{TP^2}{TP' \cdot TQ} = \frac{CD^2}{CE^2},$$

where CD, CE are the semidiameters of the ellipse parallel to TP, TQ.

Therefore $CD = CE$, and CD, CE therefore make equal angles with the axes of the ellipse; as also TP, TQ.

Now let P' approach to coincidence with P; the circle becomes the circle of curvature at P, and TQ becomes PV, the common chord of the ellipse and its circle of curvature, which therefore is equally inclined with the tangent PT to the axes.

Hence to construct the circle of curvature of an ellipse at P, draw the tangent PT and then the chord PV equally inclined with PT to the axes; PV will be the common chord of the ellipse and its circle of curvature; draw VU at right angles to PV to meet the normal at P in U, then PU will be the diameter of curvature at P.

The same construction holds for the parabola and hyperbola.

Also $$PV = \mathrm{lt}\ TQ = \mathrm{lt}\frac{TP^2}{TP'},$$

a relation which gives the chord of curvature PV in any direction and for any curve.

98. *Evolute of the Parabola.*

Let x, y be the co-ordinates of a point P on the parabola $y^2 = 2lx$; and let α, β be the co-ordinates of Q, the centre of curvature at P; to determine the locus of Q, the evolute of the parabola (fig. 30).

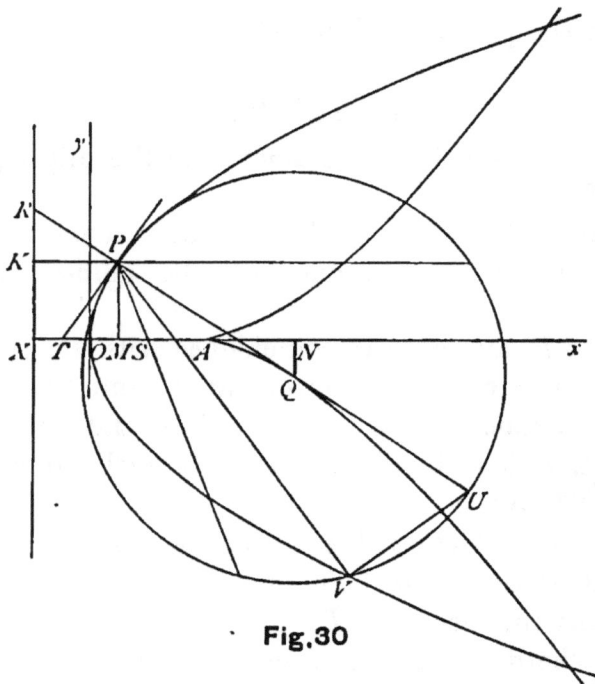

Fig. 30

Differentiating the equation of the parabola with respect to x,

$$2y\frac{dy}{dx} = 2l, \text{ or } \frac{dy}{dx} = \frac{l}{y};$$

and differentiating again

$$\frac{d^2y}{dx^2} = -\frac{l}{y^2}\frac{dy}{dx} = -\frac{l^2}{y^3}.$$

Therefore (§ 91)

$$x-a=\frac{\frac{l}{y}\left(1+\frac{l^2}{y^2}\right)}{-\frac{l^2}{y^3}}=-\frac{y^2}{l}-l=-2x-l \; ;$$

$$a=l+3x \; ;$$

and

$$y-\beta=-\frac{1+\frac{l^2}{y^2}}{-\frac{l^2}{y^3}}=\frac{y^3}{l^2}+y,$$

$$\beta=-\frac{y^3}{l^2}.$$

Conversely $x=\frac{1}{3}(a-l), \; y=-(l^2\beta)^{\frac{1}{3}} \; ;$
and since $y^2=2lx,$
therefore $l^{\frac{4}{3}}\beta^{\frac{2}{3}}=\frac{2}{3}l(a-l),$

$$l\beta^2=\frac{8}{27}(a-l)^3,$$

the equation of the evolute, which is therefore a semi-cubical parabola (§ 61).

Also

$$\rho=\frac{\left(1+\frac{l^2}{y^2}\right)^{\frac{3}{2}}}{-\frac{l^2}{y^3}}=-\frac{(l^2+y^2)^{\frac{3}{2}}}{l^2},$$

the negative sign appearing because $\dfrac{d^2y}{dx^2}$ is negative;
therefore, changing the sign,

$$PQ=\frac{(l^2+y^2)^{\frac{3}{2}}}{l^2}=\frac{PG^3}{l^2},$$

the normal at P meeting the axis of the parabola in G.

As an exercise, prove that the radius of curvature at P is $2PR$, where R is the point where the normal at P meets the directrix of the parabola.

Prove also that the evolute cuts the parabola at points which are the centres of curvature at $x=l$ of the parabola.

The arc of the evolute

$$AQ=PQ-OA=\frac{(y^2+l^2)^{\frac{3}{2}}}{l^2}-l,$$

which can also be expressed in terms of x, or a, or β.

This explains why the semi-cubical parabola can be rectified (§ 61), and in fact this was the first curve of which the rectification was effected algebraically.

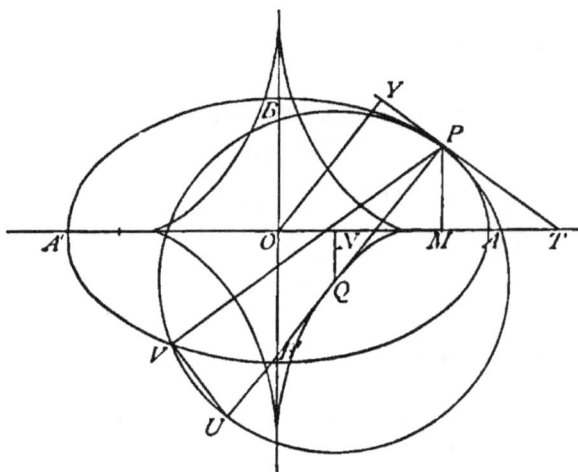

Fig.31

99. *Evolute of the Ellipse and Hyperbola.*

Take θ, the excentric angle, as the independent variable in the ellipse (§ 55), then

$$x= a\cos\theta, \qquad y= b\sin\theta;$$

$$\frac{dx}{d\theta}=-a\sin\theta, \qquad \frac{dy}{d\theta}= b\cos\theta;$$

$$\frac{d^2x}{d\theta^2}=-a\cos\theta, \qquad \frac{d^2y}{d\theta^2}=-b\sin\theta.$$

Then
$$\rho = \frac{(a^2\sin^2\theta + b^2\cos^2\theta)^{\frac{3}{2}}}{ab}$$

$$= a^2b^2\left(\frac{x^2}{a^4} + \frac{y^2}{b^4}\right)^{\frac{3}{2}} = \frac{a^2b^2}{p^3},$$

if p denotes the length of the perpendicular OY drawn from the centre O on the tangent at P (fig. 31).

Denoting the angle AOY by ω, then

$$\tan \omega = -\frac{dx}{dy} = \frac{a}{b}\tan \theta;$$

$$\sin \omega = \frac{a\sin \theta}{\sqrt{(a^2\sin^2\theta + b^2\cos^2\theta)}}, \quad \cos \omega = \frac{b\cos \theta}{\sqrt{(a^2\sin^2\theta + b^2\cos^2\theta)}}.$$

Therefore

$$x - a = \rho \cos \omega = \frac{a^2\sin^2\theta + b^2\cos^2\theta}{a}\cos \theta$$

$$= a\cos \theta - \frac{a^2 - b^2}{a}\cos^3\theta,$$

$$a = \frac{a^2 - b^2}{a}\cos^3\theta = \frac{a^2 - b^2}{a^4}x^3;$$

$$y - \beta = \rho \sin \omega = \frac{a^2\sin^2\theta + b^2\cos^2\theta}{b}\sin \theta$$

$$= b\sin \theta + \frac{a^2 - b^2}{b}\sin^3\theta,$$

$$\beta = -\frac{a^2 - b^2}{b}\sin^3\theta = -\frac{a^2 - b^2}{b^4}y^3.$$

Therefore $a\alpha = (a^2 - b^2)\cos^3\theta$, $b\beta = -(a^2 - b^2)\sin^3\theta$; and eliminating θ,

$$(a\alpha)^{\frac{2}{3}} + (b\beta)^{\frac{2}{3}} = (a^2 - b^2)^{\frac{2}{3}},$$

the equation of the evolute of the ellipse.

Similarly the equation of the evolute of the hyperbola

$$\frac{x^2}{a^2} - \frac{y^2}{b^2} = 1,$$

is
$$(a\alpha)^{\frac{2}{3}} - (b\beta)^{\frac{2}{3}} = (a^2 + b^2)^{\frac{2}{3}}.$$

As an exercise, prove that the radius of curvature at P of the ellipse or hyperbola, as well as of the parabola, is $\dfrac{PG^3}{l^2}$, where l is the semi-latus rectum and G the point where the normal meets the axis of x.

Employing the variables ξ, η of § 29, then for the ellipse,

$$a = c\frac{\cos^3\xi}{\cosh\eta}, \quad \beta = -c\frac{\sin^3\xi}{\sinh\eta}.$$

Now the equation of the tangent of the ellipse being

$$x\frac{\cos\xi}{\cosh\eta} + y\frac{\sin\xi}{\sinh\eta} = c,$$

and the equation of the confocal hyperbola being

$$\frac{x^2}{\cos^2\xi} - \frac{y^2}{\sin^2\xi} = c^2;$$

therefore (a, β) is the pole of the tangent of the ellipse with respect to the confocal hyperbola.

Hence to find Q, the centre of curvature of the ellipse at P, produce the tangent of the ellipse to meet the confocal hyperbola in R, and bisect PR in V; VO produced will meet the normal to the ellipse at P in Q.

A similar construction will give the centre of curvature at P of the confocal hyperbola.

100. *Envelopes.*

Since the normal to a curve is a tangent to the evolute (§ 92), the evolute is called the *envelope* of the normals, and the equation of the evolute is usually most readily determined from this consideration, since the point of intersection of the normal at P with a consecutive normal is ultimately Q, the point of contact of QP on the evolute (fig. 26), and also the centre of curvature at P.

For instance, the equation of the normal to the ellipse at a point whose excentric angle is θ is

$$ax \sec \theta - by \operatorname{cosec} \theta - a^2 + b^2 = 0 \dots\dots\dots\dots(\text{i.})$$

Denoting this equation by $F\theta = 0$, ·then, as explained below, to find the point of ultimate intersection of this normal with the consecutive normal, we must determine x and y from the equations

$$F\theta = 0, \text{ and } F'\theta = 0 ;$$

and the equation of the evolute is obtained by eliminating θ between these equations.

Here $F'\theta = ax \sec \theta \tan \theta + by \operatorname{cosec} \theta \cot \theta = 0 \dots\dots(\text{ii.})$; and from (i.) and (ii.),

$$ax = (a^2 - b^2)\cos^3\theta, \ by = -(a^2 - b^2)\sin^3\theta ;$$

giving the co-ordinates of the centre of curvature; and eliminating θ,

$$(ax)^{\frac{2}{3}} + (by)^{\frac{2}{3}} = (a^2 - b^2)^{\frac{2}{3}},$$

the equation of the evolute (§ 99).

Generally if $F\theta = 0$ denotes the equation of any curve, involving x and y and a *parameter* θ, as it is called ; then by varying θ a series of curves is obtained; keeping θ constant, a particular curve of the series is obtained.

To find the points of ultimate intersection of the curve $F\theta = 0$ with a consecutive curve of the series, suppose θ to receive a small increment $\Delta\theta$; then we must find x and y from the equations

$$F\theta = 0, \text{ and } F(\theta + \Delta\theta) = 0,$$

where $\Delta\theta = 0$, ultimately ;
or, which is the same thing, from the equations

$$\cdot F\theta = 0, \text{ and } \operatorname{lt} \frac{F(\theta + \Delta\theta) - F\theta}{\Delta\theta} = 0, \text{ or } F'\theta = 0.$$

The locus of these points is called the *envelope* of the series of curves, and its equation is found by eliminating θ between the equations

$$F\theta = 0, \text{ and } F'\theta = 0.$$

This curve is called the *envelope* of the series, because each curve touches the envelope where it intersects the consecutive curve of the series.

For, take three consecutive curves of the series defined by the parameters $\theta + \Delta\theta$, θ, and $\theta - \Delta\theta$; and suppose the first and second to intersect in P, and the second and third in P'; then P, P' are two consecutive points on the *envelope*, and also on the curve $F\theta = 0$, and therefore ultimately the envelope and a curve of the series have the same tangent where they meet.

Familiar instances of envelopes of straight lines, besides evolutes, are seen in *Optics* with *caustic curves*, the envelopes of rays reflected or refracted at given curves or surfaces.

Thus: (i.) The caustic of rays reflected by a circle, emanating from a point in the circumference of the circle, is a cardioid (ex. 8, p. 96);

(ii.) the caustic of rays refracted by a plane, emanating from a point, is the surface formed by the revolution of the evolute of an ellipse or hyperbola.

Examples.—(1) Find the envelope of a straight line of given length c, which moves with its ends on the co-ordinate axes.

If the straight line makes an angle θ with the axis of x, its equation is

$$\frac{x}{c\cos\theta}+\frac{y}{c\sin\theta}=1,$$

or $$x\sec\theta+y\cosec\theta=c.$$

Differentiating with respect to θ,

$$x\sec\theta\tan\theta-y\cosec\theta\cot\theta=0,$$

and therefore $x=c\cos^3\theta,\ y=c\sin^3\theta$;

and eliminating θ,

$$x^{\frac{2}{3}}+y^{\frac{2}{3}}=c^{\frac{2}{3}}.$$

(2) Find the envelope of the parabolas of § 84.

Supposing θ the variable elevation, then

$$y=x\tan\theta-\frac{gx^2}{2V^2\cos^2\theta},$$

and therefore the envelope is

$$y=\frac{V^2}{2g}-\frac{gx^2}{2V^2};$$

the equation of the parabola HQ (fig. 23).

(3) Prove that the envelope of the ellipses (§ 95)

$$\frac{x^2}{a^2}-2\frac{xy}{ab}\sin\theta+\frac{y^2}{b^2}=\cos^2\theta,$$

is $$x=\pm a,\ \text{or}\ y=\pm b.$$

(4) Find the envelope of the ellipses

$$\frac{x^2}{a^2}+\frac{y^2}{b^2}=1,\ \text{where}\ a+b=c,\ \text{a constant.}$$

Supposing a and b to be functions of some independent variable t, then to find the envelope, differentiating with respect to t, treating a and b as variable,

$$\frac{x^2}{a^3}\frac{da}{dt}+\frac{y^2}{b^3}\frac{db}{dt}=0,\ \text{and}\ \frac{da}{dt}+\frac{db}{dt}=0.$$

Multiplying the second equation by some *undetermined multiplier* λ, and adding it to the first equation,

$$\left(\frac{x^2}{a^3}+\lambda\right)\frac{da}{dt}+\left(\frac{y^2}{b^3}+\lambda\right)\frac{db}{dt}=0.$$

Now suppose λ chosen, so that

$$\frac{x^2}{a^3}+\lambda=0, \text{ then } \frac{y^2}{b^3}+\lambda=0;$$

and therefore

$$\frac{x^2}{a^2}+\frac{y^2}{b^2}+\lambda(a+b)=0,$$

or

$$1+\lambda c=0, \ \lambda=-\frac{1}{c}.$$

Therefore $\qquad a^3=cx^2, \ b^3=cy^2$;

and since $\qquad\qquad a+b=c,$

therefore $\qquad\qquad x^{\frac{2}{3}}+y^{\frac{2}{3}}=c^{\frac{2}{3}},$

the equation of the envelope.

(5) Prove that the envelope of the curves

$\left(\dfrac{x}{a}\right)^m+\left(\dfrac{y}{b}\right)^m=1$, where $a^n+b^n=c^n$, is $x^{\frac{mn}{m+n}}+y^{\frac{mn}{m+n}}=c^{\frac{mn}{m+n}}$.

101. *Maxima and Minima.*

One of the most useful applications of the Calculus is the determination of the *maximum* or *minimum* value of a function y or fx of a variable quantity x.

Suppose the curve $y=fx$ is drawn (fig. 32); then an inspection of the figure shows that the tangent to the curve is parallel to the axis of x, and therefore $\dfrac{dy}{dx}$ or $f'x=0$, at points on the curve where y has a *maximum* or *minimum* value.

The roots of the equation $f'x=0$ will be the *abscissæ* of these points, and will give the values of x which makes y a maximum or minimum.

To distinguish whether the corresponding value of y is a maximum or minimum, we notice that as x increases, $\dfrac{dy}{dx}$ or $\tan\psi$ is diminishing when y is a *maximum*, and then $\dfrac{d^2y}{dx^2}$ is *negative;* but $\dfrac{dy}{dx}$ is increasing when y is a *minimum,* and then $\dfrac{d^2y}{dx^2}$ is positive.

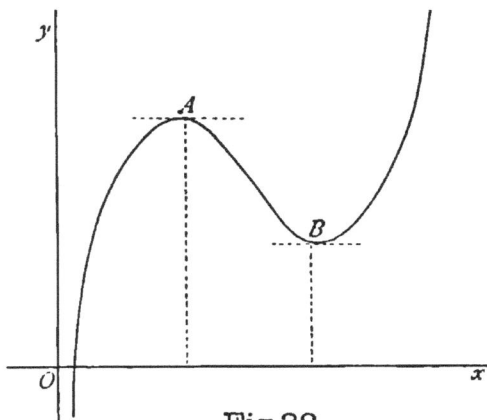

Fig.32.

This can easily be seen, with different letters, from dynamical considerations; for the velocity $\dfrac{dx}{dt}$ is positive when x is increasing, but $\dfrac{dx}{dt}$ is negative when x is diminishing; also the acceleration $\dfrac{d^2x}{dt^2}$ is positive when the velocity $\dfrac{dx}{dt}$ is increasing, but $\dfrac{d^2x}{dt^2}$ is negative when $\dfrac{dx}{dt}$ is diminishing.

As x increases continuously, the maximum and minimum values of y must occur alternately, because y after reaching a maximum value must diminish to a minimum value before increasing again to a maximum.

Examples.—(1) If $y = 2x^3 - 9x^2 + 12x - 3$, the equation of the curve of fig. 32;

$$\frac{dy}{dx} = 6x^2 - 18x + 12 = 6(x-1)(x-2) = 0,$$

when $x = 1$, or 2; and

$$\frac{d^2y}{dx^2} = 6(x-2) + 6(x-1).$$

(i.) When $x = 1$, $\frac{d^2y}{dx^2} = -6$, and $y = 2$, a maximum;

(ii.) when $x = 2$, $\frac{d^2y}{dx^2} = 6$, and $y = 1$, a minimum.

(2) If $y = x - x^2$, when $x = \frac{1}{2}$, $y = \frac{1}{4}$, a maximum.

(3) Prove that $y = x^3 - 3x^2 + 6x$ has no maximum or minimum value.

(4) If $y = (1 + x^2)(7 - x^3)^2$, when $x = 0$, $y = 49$, a minimum; $x = 1$, $y = 72$, a maximum; $x = \sqrt[3]{7}$, $y = 0$, a minimum.

(5) If $\qquad y = \dfrac{(x-1)(x-2)}{x-3}$,

when $\qquad x = 3 - \sqrt{2}$, y is a maximum;

when $\qquad x = 3 + \sqrt{2}$, y is a minimum.

(6) If $\qquad y = \dfrac{x^2 - x + 1}{x^2 + x - 1}$,

when $\qquad x = 0$, $y = -1$, a maximum;

when $\qquad x = 2$, $y = \frac{3}{5}$, a minimum.

(7) Determine the maximum and minimum values of

$$y = \frac{ax^2 + 2bx + c}{Ax^2 + 2Bx + C}.$$

This can be done algebraically, by solving this equation as a quadratic in x, and determining the limits of y from a consideration of the expression under the radical.

(8) Determine the maximum and minimum values of y, when x and y are connected by the implicit relation (§ 15)

$$x^3 - 3axy + y^3 = 0. \ .$$

Forming the first derived equation

$$3x^2 - 3ay - 3ax\frac{dy}{dx} + 3y^2\frac{dy}{dx} = 0;$$

then $\qquad \dfrac{dy}{dx} = 0$, if $x^2 - ay = 0$.

Combining this with the implicit relation we obtain

$$x = a\sqrt[3]{2}, \ y = a\sqrt[3]{4}.$$

To find the corresponding value of $\dfrac{d^2y}{dx^2}$, form the *second* derived equation, but omitting terms involving $\dfrac{dy}{dx}$, since they vanish; therefore

$$6x - 3ax\frac{d^2y}{dx^2} + 3y^2\frac{d^2y}{dx^2} = 0$$

or

$$\frac{d^2y}{dx^2} = \frac{2x}{ax - y^2} = -\frac{2}{a},$$

so that the corresponding value of y is a maximum.

Similarly, by differentiating with respect to y, equating $\dfrac{dx}{dy}$ to zero, and examining the sign of $\dfrac{d^2x}{dy^2}$, we find that when $y = a\sqrt[3]{2}$, x has the maximum value $a\sqrt[3]{4}$.

These considerations are sometimes useful in drawing a curve whose equation is given as above.

(9) If $xy(x-y) = 2$, determine the maximum and minimum values of x and y; also if

$$x^4 - 4xy + y^4 = 0.$$

M

(10) Determine the greatest rectangle which can be inscribed in a given isosceles triangle.

Let ABB' be the given isosceles triangle (fig. 33) and let a denote the altitude OA and $2b$ the base BB'; and x the height and $2y$ the breadth of the inscribed rectangle $PNN'P'$.

Then
$$\frac{x}{a} + \frac{y}{b} = 1,$$

since P lies on the straight line AB.

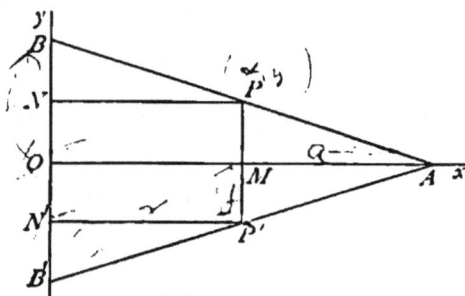

Fig.33.

Also, if u denotes the area of the rectangle,

$$u = 2xy = 2xb\left(1 - \frac{x}{a}\right) = 2b\left(x - \frac{x^2}{a}\right),$$

and
$$\frac{du}{dx} = 2b\left(1 - 2\frac{x}{a}\right) = 0,$$

when $x = \frac{1}{2}a$, $y = \frac{1}{2}b$; and then $\frac{d^2u}{dx^2} = -4\frac{b}{a}$;

so that $u = \frac{1}{2}ab$, a maximum.

(11) Determine the greatest cylinder which can be inscribed in a given cone.

Suppose the preceding figure to be made to revolve round the axis of x, and describe a cone of altitude a

and radius of base b; also a cylinder of altitude x, and radius of base y.

Then the volume of the cylinder

$$V = \pi x y^2 = \pi b^2 x \left(1 - \frac{x}{a}\right)^2 ;$$

and

$$\frac{dV}{dx} = \pi b^2 \left\{ \left(1 - \frac{x}{a}\right)^2 - 2\frac{x}{a}\left(1 - \frac{x}{a}\right) \right\}$$

$$= \pi b^2 \left(1 - \frac{x}{a}\right)\left(1 - 3\frac{x}{a}\right) = 0,$$

when $\qquad x = a$, or $x = \frac{1}{3}a$.

(i.) When $x = a$, $\dfrac{d^2V}{dx^2} = 2\pi \dfrac{b^2}{a}$; and $V = 0$, a minimum;

(ii.) When $x = \frac{1}{3}a$, $\dfrac{d^2V}{dx^2} = -2\pi \dfrac{b^2}{a}$; and

$$V = \frac{4}{27}\pi a b^2 = \frac{4}{9} \text{ volume of the cone, a maximum.}$$

(12) Determine the cylinder of greatest curved surface which can be inscribed in a given cone.

With the same notation as before, the surface $S = 2\pi x y$, and therefore, as in Ex. 10, S is a maximum when

$$x = \tfrac{1}{2}a, \quad y = \tfrac{1}{2}b.$$

(13) Determine the greatest cylinder which can be inscribed in a given sphere.

Here the volume $V = 2\pi x y^2$, where $y^2 = a^2 - x^2$; so that

$$V = 2\pi(a^2 x - x^3) ;$$

and

$$\frac{dV}{dx} = 2\pi(a^2 - 3x^2) = 0,$$

when $\qquad x^2 = \frac{1}{3}a^2$, $x = \frac{1}{3}a\sqrt{3}$; and then

$$V = \tfrac{4}{9}\sqrt{3}\pi a^3 = \tfrac{1}{3}\sqrt{3} \text{ volume of the sphere.}$$

(14) Determine the cylinder of greatest curved surface which can be inscribed in a given sphere.

Here the surface

$$S = 4\pi xy = 4\pi x \sqrt{(a^2 - x^2)}.$$

We can rationalize S, and omit constant factors, and determine the maximum value of this quantity.

Let $\qquad u = x^2(a^2 - x^2),$

then $\qquad \dfrac{du}{dx} = 2a^2x - 4x^3 = 0,$

when $\qquad x^2 = \tfrac{1}{2}a^2,\ x = \tfrac{1}{2}a\sqrt{2}\,;$

and then $\qquad \dfrac{d^2u}{dx^2} = -4a^2\,;$

so that u and therefore S is a maximum when $x = \tfrac{1}{2}a\sqrt{2}$, and then $S = 2\pi a^2 = \tfrac{1}{2}$ surface of the sphere.

(15) Prove that the volume of the greatest cone which can be inscribed in a given sphere is $\tfrac{8}{27}$ of the volume of the sphere.

(16) Determine the proportions of a cylinder of given volume open at one end, in order that the surface should be a minimum.

(The length equal to the radius of the cylinder.)

(17) Determine the proportions of a cylinder of given volume, closed at both ends, in order that the whole surface should be a minimum.

(The length equal to the diameter.)

(18) Determine the proportions of a cylindrical tin canister to have a maximum volume for a given amount of metal, supposing the ends doubled down to overlap cylindrically (i.) a given distance, (ii.) a given fraction of the length of the cylinder.

(The diameter equal to (i.) the difference, (ii.) the sum of the lengths of the cylinder and of the ends.)

(19) Prove that in a conical cup, of given surface and holding the maximum quantity of water, the height is $\sqrt{2}$ times the radius of the rim.

(20) Prove that, according to the regulations of the Parcel Post, which require the sum of the length and girth of a parcel not to exceed 6 feet—

(i.) The greatest sphere allowed is about $17\frac{3}{8}$ inches in diameter, and a little over $1\frac{1}{2}$ cubic feet in volume;

(ii.) The greatest cube is $14\frac{2}{3}$ inches long, and nearly $1\frac{3}{4}$ cubic feet in volume;

(iii.) The greatest rectangular box is 2 feet long and 1 foot square, and 2 cubic feet in volume;

(iv.) The greatest parcel of any shape is a cylinder, 2 feet long, and 4 feet in girth, and over $2\frac{1}{2}$ cubic feet in volume. (*Rev. W. A. Whitworth.*)

(21) Determine the most economical speed in fuel of a steamer against a tide, supposing the resistance to vary as the nth power of the velocity through the water.

Let a denote the velocity of the tide, x the velocity of the steamer through the water; then $x-a$ will be the velocity of the steamer relatively to the land.

The power required and therefore the coal burnt per hour will vary as the product of the resistance and the speed, that is, as x^{n+1}, and therefore the coal burnt per mile will vary as

$$\frac{x^{n+1}}{x-a}.$$

By the ordinary rule, this is a minimum when

$$x = \frac{n+1}{n}a.$$

102. In finding maxima and minima, exceptional cases sometimes occur, where for a certain value of x, not only $\frac{dy}{dx} = 0$, but also $\frac{d^2y}{dx^2} = 0$, $\frac{d^3y}{dx^3} = 0$, ...

In such cases it is generally simpler to notice that if x increases continuously, $\frac{dy}{dx}$ changes sign from positive to negative as y passes through a maximum value, and $\frac{dy}{dx}$ changes sign from negative to positive as y passes through a minimum value; but if $\frac{dy}{dx}$ does not change sign, y is neither a maximum or minimum.

Sometimes also y has a maximum or minimum value when $\frac{dy}{dx}$ changes sign by passing through the value infinity; but these cases require special investigation, and are conveniently solved by tracing the curve whose equation is $y = fx$.

These cases are represented graphically in fig. 34.

At A, y has a maximum, and at B, a minimum value.

At C, $\frac{dy}{dx} = 0$, but does not change sign, so that y is neither a maximum or minimum, and C is called a *point of inflexion* on the curve.

At D, $\frac{dy}{dx} = \infty$, and changes sign from positive to negative, so that y is a maximum, and D is called a *cusp*.

At E, $y = \infty$, and $\frac{dy}{dx} = \infty$, but does not change sign, and y changes sign from $-\infty$ to $+\infty$ on crossing the *asymptote*.

At F, $y = \infty$, and $\dfrac{dy}{dx} = \infty$, and changes sign from positive to negative, and y has an infinite maximum value.

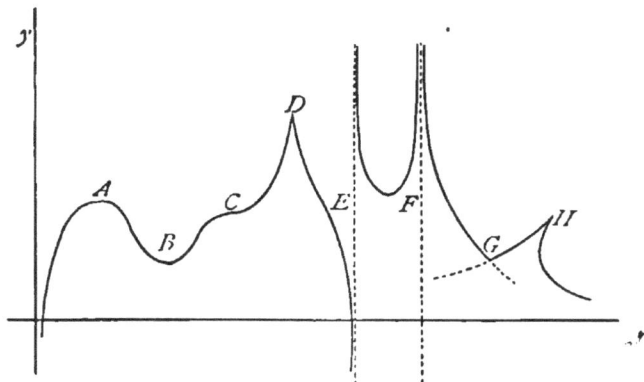

Fig.34.

At G, $\dfrac{dy}{dx}$ is discontinuous, and changes abruptly from a negative value to a positive value, and y is a minimum.

At H, y has a maximum value, but $\dfrac{dy}{dx}$ does not vanish. (De Morgan, *Diff. and Int. Calculus*, p. 45.)

Examples.—(1) If $y = x^5 - 5x^4 + 5x^3$,

then $\qquad \dfrac{dy}{dx} = 5x^4 - 20x^3 + 15x^2 = 5x^2(x-1)(x-3) = 0$,

when $x = 0$, or 1, or 3.

When $x = 0$, $\dfrac{d^2y}{dx^2} = 0$, and $\dfrac{dy}{dx}$ does not change sign, so that y is neither a maximum or minimum, as at C.

When $x = 1$, $\dfrac{d^2y}{dx^2} = -10$, and $y = 1$, a maximum; and

when $x = 3$, $\dfrac{d^2y}{dx^2} = 90$, and $y = -27$, a minimum.

(2) If $\dfrac{dy}{dx} = (x-1)(x-2)^2(x-3)^3$,

then $\dfrac{dy}{dx} = 0$, when $x = 1$, or 2, or 3.

When $x = 1$, $\dfrac{dy}{dx}$ changes from positive to negative, and y is a maximum; when $x = 2$, $\dfrac{dy}{dx}$ does not change sign, and y is neither a maximum or minimum; similarly when $x = 3$, y is a minimum.

(3) If $y = x^m(a-x)^n$,

then $\dfrac{dy}{dx} = x^{m-1}(a-x)^{n-1}\{ma - (m+n)x\} = 0$,

when $x = 0$, or $\dfrac{ma}{m+n}$, or a.

When $x = 0$, $\dfrac{dy}{dx}$ changes sign from negative to positive, if m is even, and y is then a minimum; but $\dfrac{dy}{dx}$ does not change sign if m is odd, and then y is neither a maximum or minimum.

Similarly when $x = a$, y is a minimum if n is even; y is neither a maximum or minimum if n is odd.

Therefore the intermediate value of $x = \dfrac{ma}{m+n}$ makes y a maximum.

(4) If $y = a - b(x-c)^{\frac{2}{3}}$,

then $\dfrac{dy}{dx} = \infty$, when $x = c$, and changes from positive to negative, so that $y = a$ is a maximum, as at D (fig. 34).

(5) If $y = (x^{\frac{3}{2}}+1)(x-7)^2$,

$\dfrac{dy}{dx} = \tfrac{2}{3}x^{-\frac{1}{3}}(x^{\frac{1}{3}}-1)(x-7)(4x^{\frac{3}{2}}+4x^{\frac{1}{3}}+7)$.

When $x = 0$, $\dfrac{dy}{dx} = \infty$, and changes from negative to positive, so that y is a minimum.

When $x = 1$, $\dfrac{dy}{dx} = 0$, and y is a maximum; and when $x = 7$, y is a minimum.

103. Some maximum and minimum problems, which would be very complicated solved otherwise, can be solved very simply from the consideration that the successive values of a function are equal when the function is a maximum or minimum.

Thus to determine the maximum or minimum area cut off from a conic by a normal chord PR (fig. 35), suppose

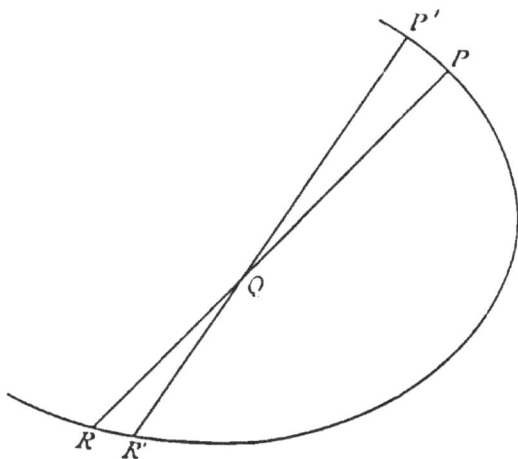

Fig.35.

PR and the consecutive normal chord $P'R'$ to cut off equal areas, then the area $PQP' =$ area RQR', if Q is the point of intersection of PR and $P'R'$; and therefore Q, which is ultimately the centre of curvature at P, is then ultimately the middle point of PR; so that PR is the diameter of curvature when it cuts off a maximum or minimum area; and then PR makes an angle $\frac{1}{4}\pi$ with an axis of the conic (§ 97).

Again, to determine the point P, such that the sum of its distances from three given points A, B, C is a minimum; suppose, firstly, that the sum of the distances from two of the given points, B and C, is given; then P is constrained to move on an ellipse of which B and C are the foci; and then the distance AP is a minimum when AP is a normal to this ellipse, and therefore makes equal angles with BP and CP.

Therefore, by symmetry, AP, BP, and CP make equal angles of 120° with each other when $AP + BP + CP$ is a minimum.

In the same way it may be proved that a triangle DEF inscribed in a triangle ABC has a minimum perimeter when FD, DE make equal angles with BC, DE, EF with CA, and EF, FD with AB; and then D, E, F are the feet of the perpendiculars drawn from A, B, C on the opposite sides BC, CA, AB of the triangle ABC.

Again, to find the maximum triangle with given base and given vertical angle, since the vertex in this case is constrained to move on the arc of a circle, the area will be a maximum when the altitude is a maximum, and the triangle is then isosceles.

Suppose it is required to determine the path APB of a ray of light from A to B which shall take the shortest time, supposing the velocity of light to change from v to v' in crossing the curve PP' (fig. 36).

Then from the consideration that the time in a consecutive path $AP'B$ is the same as in APB,

$$\frac{Pq}{v} = \frac{Pr}{v'}$$

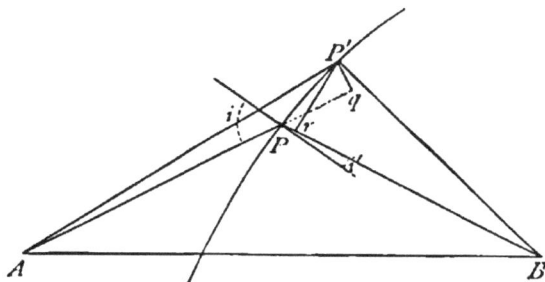

Fig. 36.

ultimately, where $P'q$, $P'r$ are the perpendiculars from P' on AP and PB; and therefore

$$\frac{v}{v'} = \mathrm{lt}\frac{Pq}{Pr} = \frac{\sin i}{\sin i''}$$

where i, i' are the angles AP, PB make with the normal to the curve PP' at P.

In this manner Fermat proved the law of the *refraction* of light.

Some geometrical problems of maximum and minimum can by *projection* (Salmon or Smith's *Conic Sections*) be made to depend on a simpler problem, the solution of which is intuitive.

Thus to find the maximum ellipse which can be inscribed in a triangle, or the minimum ellipse which can be described about a triangle, project the triangle orthogonally into an equilateral triangle; the maximum inscribed ellipse is then the inscribed circle, and the minimum circumscribed ellipse is the circumscribed circle.

The problem of finding the maximum rectangle which can be inscribed in an ellipse, or the maximum rectangular parallelepiped which can be inscribed in an ellipsoid, is thus by projection reduced to the corresponding problem for a circle and a sphere, the solution of which is a square and a cube.

Sometimes mechanical considerations are useful in determining maxima and minima; thus the celebrated problem of the shape of the cells of bees, for the greatest economy of wax, is seen to be the same problem as the arrangement of the capillary films of an aggregation of regular soap bubbles, which tend to arrange themselves so that the surface is a minimum.

The surface tension being uniform, three faces will meet in an edge at equal angles of 120°; and the corners or summits will be formed by the meeting of four or eight edges, so that the simplest and most regular elementary cell is a *rhombic dodecahedron* (*vide* a paper by Mrs. Bryant, D.Sc., read before the London Mathematical Society, March, 1885).

CHAPTER IV.

EXPANSION OF FUNCTIONS.

104. *Taylor's Theorem and Maclaurin's Theorem.*

In Algebra and Trigonometry the Binomial and Exponential Theorems are established, by means of which it is shown how to expand $(a+x)^m$, a^x, e^x, log $(1+x)$, sin x, cos x,... in ascending powers of x; and we shall now show that all these and similar expansions are particular cases of one general Theorem called *Taylor's Theorem*, by means of which any function whatever can be expanded.

Taylor's Theorem is due to Brook Taylor, and was given by him in his "*Methodus Incrementorum Directa et Inversa*" in 1715.

The Theorem asserts that if $f(a+x)$ is expanded in ascending positive integral powers of x, then

$$f(a+x) = fa + x\, f'a + \frac{x^2}{2!}f''a + \frac{x^3}{3!}f'''a + \ldots + \frac{x^n}{n!}f^n a + \ldots$$

For assume that

$$f(a+x) = A_0 + xA_1 + x^2A_2 + x^3A_3 + \ldots + x^nA_n + \ldots$$

where the A's are functions of a only, and do not contain x.

First put $\qquad x = 0$, then $fa = A_0$.

Next differentiate successively with respect to x, and put $x = 0$ after each differentiation; then

$$f'(a+x) = A_1 + 2xA_2 + \ldots\ldots\ldots, \quad f'a = \quad A_1;$$
$$f''(a+x) = 2!\,A_2 + 2\,.\,3xA_3 + \ldots\ldots, \quad f''a = 2!\,A_2;$$
$$f'''(a+x) = 3!\,A_3 + \ldots\ldots\ldots\ldots, \quad f'''a = 3!\,A_3;$$
$$\ldots\ldots\ldots\ldots\ldots\ldots\ldots\ldots\ldots$$
$$f^n(a+x) = n!\,A_n + \ldots\ldots\ldots\ldots, \quad f^n a = n!\,A_n.$$

Therefore

$$A_0 = fa,\ A_1 = f'a,\ A_2 = \frac{1}{2!}f''a,\ A_3 = \frac{1}{3!}f'''a, \text{ and generally}$$

$$A_n = \frac{1}{n!}f^n a.$$

Therefore

$$f(a+x) = fa + x\,f'a + \frac{x^2}{2!}f''a + \frac{x^3}{3!}f'''a + \ldots\ldots + \frac{x^n}{n!}f^n a + \ldots\ldots,$$

which is Taylor's Theorem.

As a simple illustration, consider the example of supposing $fa = a^m$, then $f'a = ma^{m-1}$, $f''a = m(m-1)a^{m-2}$, $\ldots\ldots$; and therefore

$$f(a+x) = (a+x)^m$$
$$= a^m + ma^{m-1}x + \frac{m(m-1)}{2!}a^{m-2}x^2 + \ldots\ldots,$$

the *Binomial Theorem*.

Symbolical Form of Taylor's Theorem.

The expansion may be written

$$f(a+x) = \left(1 + x\frac{d}{da} + \frac{x^2}{2!}\frac{d^2}{da^2} + \frac{x^3}{3!}\frac{d^3}{da^3} + \ldots + \frac{x^n}{n!}\frac{d^n}{da^n} + \ldots\right)fa,$$

and treating the operator $\frac{d}{da}$ as an algebraical quantity, this is equivalent to

$$f(a+x) = e^{x\frac{d}{da}} fa,$$

which is called the *symbolical form* of Taylor's Theorem.

Maclaurin's Theorem.

Suppose $a = 0$, then

$$fx = f0 + xf'0 + \frac{x^2}{2!}f''0 + \frac{x^3}{3!}f'''0 + \ldots\ldots + \frac{x^n}{n!}f^n0 + \ldots\ldots$$

which is called *Maclaurin's Theorem*, but this theorem was first given by Stirling in 1717.

The meaning of f^n0 is that fx must be differentiated n times with respect to x, and then x put equal to 0 *after* the differentiation.

In Taylor's Theorem $f(a+x)$ is expanded in ascending powers of x, a part of the whole argument $a+x$ of the function $f(a+x)$; in Maclaurin's Theorem fx is expanded in powers of x, the whole argument of the function of x.

We might have proved Maclaurin's Theorem first in the above manner in which Taylor's Theorem was obtained, and then have derived Taylor's Theorem by putting

$$fx = F(a+x);$$

and then differentiating n times with respect to x,

$$f^nx = F^n(a+x); \text{ and putting } x = 0, f^n0 = F^na.$$

Substituting in Maclaurin's Theorem,

$$F(a+x) = Fa + xF'a + \frac{x^2}{2!}F''a + \ldots\ldots + \frac{x^n}{n!}F^na + \ldots\ldots,$$

which is Taylor's Theorem.

In fact the two theorems of Taylor and Maclaurin are the same when considered geometrically as the equation of a curve with different origins on the axis of x.

105. *Application to the Expansion of Functions.*

(1) Let $\quad fx = \sin x$, then $f0 = 0$;

also $\qquad f^{2n}x = (-1)^n \sin x$, and $f^{2n}\,0 = 0$;

$\qquad f^{2n+1}x = (-1)^n \cos x$, and $f^{2n+1}0 = (-1)^n$.

Therefore, by Maclaurin's Theorem,

$$\sin x = x - \frac{x^3}{3!} + \frac{x^5}{5!} - \ldots\ldots + \frac{(-1)^n x^{2n+1}}{(2n+1)!} + \ldots\ldots$$

(2) Let $\quad fx = \cos x$, then $f0 = 1$;

$\qquad f^{2n}x = \quad (-1)^n \cos x$, and $f^{2n}\,0 = (-1)$;

$\qquad f^{2n+1}x = -(-1)^n \sin x$, and $f^{2n+1}0 = 0$.

Therefore, by Maclaurin,

$$\cos x = 1 - \frac{x^2}{2!} + \frac{x^4}{4!} - \ldots\ldots + \frac{(-1)^n x^{2n}}{2n!} + \ldots\ldots$$

(3) Let $\quad fx = e^x$, then $f0 = 1$;

also $\qquad f^n x = e^x$, and $f^n 0 = 1$.

Therefore

$$e^x = 1 + x + \frac{x^2}{2!} + \ldots\ldots + \frac{x^n}{n!} + \ldots\ldots$$

Changing x into cx,

$$e^{cx} = 1 + cx + \frac{c^2 x^2}{2!} + \ldots\ldots + \frac{c^n x^n}{n!} + \ldots\ldots$$

Again, suppose

$\qquad fx = a^x, \qquad\qquad$ then $f0 = 1$;

$\qquad f^n x = a^x (\log a)^n$, and $f^n 0 = (\log a)^n$.

Therefore

$$a^x = 1 + x \log a + \frac{x^2 (\log a)^2}{2!} + \ldots\ldots + \frac{x^n (\log a)^n}{n!} + \ldots\ldots$$

This expansion is the same as the preceding, if $c = \log a$, since $a^x = e^{x \log a}$.

(4) $\qquad \sinh x = \tfrac{1}{2}(e^x - e^{-x})$

$$= x + \frac{x^3}{3!} + \frac{x^5}{5!} + \ldots\ldots + \frac{x^{2n+1}}{(2n+1)!} + \ldots\ldots$$

(5) $\cosh x = \frac{1}{2}(e^x + e^{-x})$

$$= 1 + \frac{x^2}{2!} + \frac{x^4}{4!} + \ldots\ldots + \frac{x^{2n}}{2n!} + \ldots\ldots$$

These expansions might have been obtained independently by Maclaurin's Theorem.

(6) Let

$$f(1+x) = \log(1+x), \text{ then } f1 = 0;$$

$$f'(1+x) = \frac{1}{1+x}, \qquad f'1 = 1;$$

$$f''(1+x) = -\frac{1}{(1+x)^2}, \qquad f''1 = -1;$$

$$\ldots\ldots\ldots\ldots\ldots\ldots\ldots\ldots\ldots\ldots\ldots$$

$$f^n(1+x) = \frac{(-1)^{n-1}(n-1)!}{(1+x)^n}, \ f^n1 = (-1)^{n-1}(n-1)!.$$

Therefore by Taylor's Theorem (here $a = 1$)

$$\log(1+x) = x - \frac{x^2}{2} + \frac{x^3}{3} - \frac{x^4}{4} + \ldots\ldots + \frac{(-1)^{n-1}x^n}{n} + \ldots\ldots$$

We cannot expand $\log x$ in ascending powers of x, because if $fx = \log x$, then $f0$, $f'0$, $f''0$, ... are all infinite; the same remark applies to $\exp\left(-\frac{1}{x}\right)$, $\exp\left(-\frac{1}{x^2}\right)\ldots$

(7) Then from the preceding expansion,

$$\tanh^{-1}x = \frac{1}{2}\log\frac{1+x}{1-x} = x + \frac{x^3}{3} + \frac{x^5}{5} + \ldots\ldots + \frac{x^{2n+1}}{2n+1} + \ldots\ldots$$

106. Some expansions can be derived from others by differentiation or integration; thus

$$\cos x = \frac{d\sin x}{dx}, \ \cosh x = \frac{d\sinh x}{dx},$$

giving the expansion of $\cos x$ or $\cosh x$ when that of $\sin x$ or $\sinh x$ is known.

N

Again, by integration,

$$\log(1+x)=\int_0^{}\frac{dx}{1+x}=\int_0^{}(1+x)^{-1}dx$$

$$=\int_0^{}(1-x+x^2-x^3+...)dx=x-\frac{x^2}{2}+\frac{x^3}{3}-\frac{x^4}{4}+...$$

And $\quad \tanh^{-1}\frac{x}{a}=\int_0^{}\frac{adx}{a^2-x^2}$

$$=\int_0^{}\left(\frac{1}{a}+\frac{x^2}{a^3}+\frac{x^4}{a^5}+......\right)dx=\frac{x}{a}+\frac{x^3}{3a^3}+\frac{x^5}{5a^5}+......$$

Also $\quad \tan^{-1}\frac{x}{a}=\int_0^{}\frac{adx}{a^2+x^2}$

$$=\int_0^{}\left(\frac{1}{a}-\frac{x^2}{a^3}+\frac{x^4}{a^5}-......,\right)dx=\frac{x}{a}-\frac{x^3}{3a^3}+\frac{x^5}{5a^5}-...,$$

which is called *Gregorie's series.*

Similarly, $\quad \sin^{-1}\frac{x}{a}=\int_0^{}\frac{dx}{\sqrt{(a^2-x^2)}}=\frac{1}{a}\int_0^{}\left(1-\frac{x^2}{a^2}\right)^{-\frac{1}{2}}dx$

$$=\int_0^{}\left(1+\frac{1}{2}\frac{x^2}{a^2}+\frac{1.3}{2.4}\frac{x^4}{a^4}+\frac{1.3.5}{2.4.6}\frac{x^6}{a^6}+......\right)d\frac{x}{a}$$

$$=\frac{x}{a}+\frac{1}{2}\frac{x^3}{3a^3}+\frac{1.3}{2.4}\frac{x^5}{5a^5}+\frac{1.3.5}{2.4.6}\frac{x^7}{7a^7}+......;$$

and $\quad \sinh^{-1}\frac{x}{a}=\int_0^{}\frac{dx}{\sqrt{(a^2+x^2)}}=\frac{x}{a}-\frac{1}{2}\frac{x^3}{3a^3}+\frac{1.3}{2.4}\frac{x^5}{5a^5}-......$

To expand a rational algebraical function, we may resolve it into partial fractions, (§ 39), and expand each of these in ascending powers of x by the Binomial Theorem.

When only a few terms at the beginning of a series are required, they may often be readily obtained by a combination of the expansion of the constituents of the function.

Examples.—(1) Find the first four terms in the expansion in ascending powers of x of (i.) $\dfrac{x}{(x-a)(x-b)(x-c)}$,
(ii.) $e^x \sec x$, (iii.) $\exp(a \tan^{-1}x)$, (iv.) $e^{\frac{1}{2}x^2}\sin x$, (v.) $e^{\frac{1}{2}x^2}\cos x$,
(vi.) $(\cos x)^n$; (vii.) $(\cosh x)^n$, (viii.) $\tan x$, (ix.) $\sec x$,
(x.) $x \cot x$, (xi.) $x \operatorname{cosec} x$, (xii.) $(1+x)^x$.

For instance (vii.),

$$
\begin{aligned}
(\cosh x)^n &= \left(1 + \frac{x^2}{2} + \frac{x^4}{24} + \frac{x^6}{720} + \ldots\ldots\right)^n \\
&= 1 + n\left(\frac{x^2}{2} + \frac{x^4}{24} + \frac{x^6}{720} + \ldots\ldots\right) \\
&\quad + \frac{n(n-1)}{2}\left(\frac{x^2}{2} + \frac{x^4}{24} + \ldots\ldots\right)^2 \\
&\quad + \frac{n(n-1)(n-2)}{6}\left(\frac{x^2}{2} + \ldots\ldots\right)^3 + \ldots\ldots \\
&= 1 + \tfrac{1}{2}nx^2 + \frac{3n^2-2n}{24}x^4 + \frac{15n^3-30n^2+16n}{720}x^6 + \ldots\ldots
\end{aligned}
$$

(2) Prove that

$$
e^{x\cos a}\cos(x\sin a) = 1 + x\cos a + \frac{x^2}{2!}\cos 2a + \frac{x^3}{3!}\cos 3a
$$

$$
+ \ldots\ldots + \frac{x^n}{n!}\cos na + \ldots\ldots
$$

(3) $e^{ax}\cos px = 1 + ax + \dfrac{a^2x^2}{2!}\cos 2a(\sec a)^2 + \dfrac{a^3x^3}{3!}\cos 3a(\sec a)^3$

$$
+ \ldots\ldots + \frac{a^n x^n}{n!}\cos na(\sec a)^n + \ldots\ldots,
$$

where $\tan a = \dfrac{p}{a}$.

(4) $\quad \cosh ax \cos px = 1 + \dfrac{a^2x^2}{2!}\cos 2a(\sec a)^2$

$$
+ \frac{a^4x^4}{4!}\cos 4a(\sec a)^4 + \ldots + \frac{a^{2n}x^{2n}}{2n!}\cos 2na(\sec a)^{2n} + \ldots
$$

107. *Bernoulli's Numbers.*

In the expansion of $\tan x$, $\cot x$, $\operatorname{cosec} x$, $\tanh x$, $\coth x$, $\operatorname{cosech} x$, the coefficients involve certain numbers called Bernoulli's numbers, which are thus defined.

Suppose $\quad \frac{1}{2}x\dfrac{e^x+1}{e^x-1}$ or $\dfrac{x}{e^x-1}+\frac{1}{2}x$

to be expanded in ascending powers of x; then only *even* powers will occur, because the expression is an *even* function of x, being unchanged when $-x$ is written for x (§ 50).

Writing the expansion in the form

$$\tfrac{1}{2}x\frac{e^x+1}{e^x-1}=1+\frac{x^2}{2!}B_1-\frac{x^4}{4!}B_2+\ldots\ldots+(-1)^{n-1}\frac{x^{2n}}{2n!}B_n+\ldots\ldots,$$

then $B_1,\ B_2,\ \ldots,\ B_n,\ \ldots$ are called *Bernoulli's Numbers.*

With our notation (§ 28)

$$\tfrac{1}{2}x\frac{e^x+1}{e^x-1}=\tfrac{1}{2}x\coth\tfrac{1}{2}x,$$

so that, changing $\frac{1}{2}x$ into x,

$$x\coth x=1+\frac{x^2}{2!}2^2B_1-\frac{x^4}{4!}2^4B_2+\ldots+(-1)^{n-1}\frac{x^{2n}}{2n!}2^{2n}B_n+..\text{(i.).}$$

Again, changing x into ix, then (§ 28)

$$ix\coth ix=x\cot x$$

$$=1-\frac{x^2}{2!}2^2B_1-\frac{x^4}{4!}2^4B_2-\ldots-\frac{x^{2n}}{2n!}2^{2n}B_n-\ldots$$

or $\quad \cot x=\dfrac{1}{x}-\dfrac{x}{2!}2^2B_1-\dfrac{x^3}{4!}2^4B_2-\ldots-\dfrac{x^{2n-1}}{2n!}2^{2n}B_n-\ldots\ldots\text{(ii.),}$

Now $\tan x = \cot x - 2 \cot 2\,x$, so that

$$\tan x = \frac{x}{2!}\,2^2(2^2-1)B_1 + \frac{x^3}{4!}\,2^4(2^4-1)B_2 + \dots$$
$$+ \frac{x^{2n-1}}{2n!}2^{2n}(2^{2n}-1)B_n + \dots\dots\dots\dots\dots(\text{iii.}) ;$$

and therefore changing x into ix,

$$\tanh x = \frac{x}{2!}2^2(2^2-1)B_1 - \frac{x^3}{4!}2^4(2^4-1)B_2 + \dots$$
$$+ (-1)^{n-1}\frac{x^{2n-1}}{2n!}2^{2n}(2^{2n}-1)B_n + \dots\dots\dots(\text{iv.}).$$

Again,

$$\operatorname{cosec} x = \tan \tfrac{1}{2}\,x + \cot x,$$

so that

$$\operatorname{cosec} x = \frac{1}{x} + \frac{x}{2!}\,2B_1 + \frac{x^3}{4!}\,2(2^3-1)B_2 + \frac{x^5}{6!}2(2^5-1)B_3 + \dots(\text{v.}).$$

and therefore

$$\operatorname{cosech} x = \frac{1}{x} - \frac{x}{2!}\,2B_1 + \frac{x^3}{4!}\,2(2^3-1)B_2 - \frac{x^5}{6!}2(2^5-1)B_3 + \dots(\text{vi.}).$$

The first nine numbers of Bernoulli are

$$B_1 = \frac{1}{6},\ B_2 = \frac{1}{30},\ B_3 = \frac{1}{42},\ B_4 = \frac{1}{30},\ B_5 = \frac{5}{66},\ B_6 = \frac{691}{2730},$$
$$B_7 = \frac{7}{6},\ B_8 = \frac{3617}{510},\ B_9 = \frac{43867}{798}.$$

Resolving $\sin x$ into factors (ex. 94, p. 58),

$$\sin x = x\left(1 - \frac{x^2}{\pi^2}\right)\left(1 - \frac{x^2}{2^2\pi^2}\right)\left(1 - \frac{x^2}{3^2\pi^2}\right)\dots\dots,$$

and taking logarithms,

$$\log \sin x = \log x + \log\left(1 - \frac{x^2}{\pi^2}\right) + \log\left(1 - \frac{x^2}{2^2\pi^2}\right)$$
$$+ \log\left(1 - \frac{x^2}{3^2\pi^2}\right) + \dots\dots,$$

and expanding the logarithms (§ 105),

$$\log \sin x = \log x - \frac{x^2}{\pi^2}\left(1 + \frac{1}{2^2} + \frac{1}{3^2} + \ldots\ldots\right)$$
$$- \frac{x^4}{2\pi^4}\left(1 + \frac{1}{2^4} + \frac{1}{3^4} + \ldots\ldots\right)$$
$$- \ldots\ldots$$
$$- \frac{x^{2n}}{n\pi^{2n}}S_{2n} - \ldots\ldots,$$

where
$$S_p = 1 + \frac{1}{2^p} + \frac{1}{3^p} + \ldots\ldots$$

Differentiating
$$\cot x = \frac{1}{x} - \frac{2x}{\pi^2}S_2 - \frac{2x^3}{\pi^4}S_4 - \ldots - \frac{2x^{2n-1}}{\pi^{2n}}S_{2n} - \ldots\ldots,$$

so that, comparing with equation (ii.),
$$B_n = \frac{2(2n)!}{(2\pi)^{2n}}S_{2n}.$$

Similarly resolving $\cos x$ into factors, and taking logarithms

$$\log \cos x = - \frac{x^2}{(\frac{1}{2}\pi)^2}\left(1 + \frac{1}{3^2} + \frac{1}{5^2} + \ldots\right)$$
$$- \frac{x^4}{2(\frac{1}{2}\pi)^4}\left(1 + \frac{1}{3^4} + \frac{1}{5^4} + \ldots\right)$$
$$- \ldots\ldots$$
$$- \frac{x^{2n}}{n(\frac{1}{2}\pi)^{2n}}T_{2n} - \ldots\ldots$$

where
$$T_p = 1 + \frac{1}{3^p} + \frac{1}{5^p} + \ldots\ldots$$

But $S_p - T_p = \frac{1}{2^p}S_p$, so that $T_p = \frac{2^p - 1}{2^p}S_p$;

and differentiation will give the expansion of $\tan x$.

Suppose $\sec x$ expanded in ascending powers of x; then since $\sec x$ is an even function of x, only even powers of x will appear; and then if the expansion is

$$\sec x = 1 + \frac{x^2}{2!}E_2 + \frac{x^4}{4!}E_4 + \ldots\ldots \ldots\ldots\ldots (\text{vii.}),$$

$E_2, E_4,\ldots\ldots$ are called *Euler's numbers.*

If we resolve $\dfrac{1 + \sin x}{1 - \sin x}$ into factors, and differentiate logarithmically as before, then, since

$$\sec x = \frac{1}{2}\frac{d}{dx}\log\frac{1 + \sin x}{1 - \sin x},$$

we shall find

$$E_{2n} = \frac{2(2n)!}{(\frac{1}{2}\pi)^{2n+1}}\left(1 - \frac{1}{3^{2n+1}} + \frac{1}{5^{2n+1}} - \frac{1}{7^{2n+1}} + \ldots\ldots\right).$$

For, resolved into factors,

$$\frac{1 + \sin x}{1 - \sin x} = \tan^2\left(\frac{1}{4}\pi + \frac{1}{2}x\right)$$

$$= \left\{\frac{\left(1 + \frac{2x}{\pi}\right)\left(1 - \frac{2x}{3\pi}\right)\left(1 + \frac{2x}{5\pi}\right)\ldots\ldots}{\left(1 - \frac{2x}{\pi}\right)\left(1 + \frac{2x}{3\pi}\right)\left(1 - \frac{2x}{5\pi}\right)\ldots\ldots}\right\}^2,$$

and therefore

$$\frac{1}{2}\log\frac{1 + \sin x}{1 - \sin x} = \frac{2x}{\frac{1}{2}\pi}\left(1 - \frac{1}{3} + \frac{1}{5} - \frac{1}{7} + \ldots\right)$$
$$+ \frac{2x^3}{3(\frac{1}{2}\pi)^3}\left(1 - \frac{1}{3^3} + \frac{1}{5^3} - \frac{1}{7^3} + \ldots\right)$$
$$+ \frac{2x^5}{5(\frac{1}{2}\pi)^5}\left(1 - \frac{1}{3^5} + \frac{1}{5^5} - \frac{1}{7^5} + \ldots\right) + \ldots;$$

and the coefficient of x reduces to unity, because

$$1 - \frac{1}{3} + \frac{1}{5} - \frac{1}{7} + \ldots\ldots = \frac{1}{4}\pi,$$

by putting $\dfrac{x}{a} = 1$ in Gregorie's series (§ 106).

We can express S_p, T_p, B_n and E_{2n} as *definite* integrals.

For if p is a positive integer, and m is positive, then integrating by parts, the *definite* integral (§ 49)

$$\int_0^\infty e^{-mu}u^{p-1}du = \frac{p-1}{m}\int_0^\infty e^{-mu}u^{p-2}du,$$

a formula of reduction (§ 67), and therefore finally,

$$\int_0^\infty e^{-mu}u^{p-1}du = \frac{(p-1)!}{m^{p-1}}\int_0^\infty e^{-mu}du = \frac{(p-1)!}{m^p}.$$

Therefore

$$T_p = \frac{1}{(p-1)!}\int_0^\infty (e^{-u}+e^{-3u}+e^{-5u}+\ldots\ldots)u^{p-1}du$$

$$= \frac{1}{(p-1)!}\int_0^\infty \frac{e^{-u}u^{p-1}du}{1-e^{-2u}} = \frac{2}{(p-1)!}\int_0^\infty \frac{u^{p-1}du}{\sinh u}.$$

And $\quad S_p = \dfrac{2^p}{2^p-1}T_p = \dfrac{2^{p+1}}{(2^p-1)(p-1)!}\displaystyle\int_0^\infty \dfrac{u^{p-1}du}{\sinh u}.$

Then $$B_n = \frac{8n}{(2^{2n}-1)\pi^{2n}}\int_0^\infty \frac{u^{2n-1}du}{\sinh u}.$$

Similarly

$$E_{2n} = \frac{2}{(\frac{1}{2}\pi)^{2n+1}}\int_0^\infty (e^{-u}-e^{-3u}+e^{-5u}-\ldots\ldots)u^{2n}du$$

$$= \frac{2}{(\frac{1}{2}\pi)^{2n+1}}\int_0^\infty \frac{e^{-u}u^{2n}du}{1+e^{-2u}} = \frac{4}{(\frac{1}{2}\pi)^{2n+1}}\int_0^\infty \frac{u^{2n}du}{\cosh u}.$$

Put $\theta = \mathrm{gd}\, u$, then $u = \log \tan(\frac{1}{4}\pi+\frac{1}{2}\theta)$ (§ 30), and

$$E_{2n} = \frac{4}{(\frac{1}{2}\pi)^{2n+1}}\int_0^{\frac{1}{2}\pi} \{\log \tan(\tfrac{1}{4}\pi+\tfrac{1}{2}\theta)\}^{2n}d\theta.$$

Put $\quad u = \frac{1}{2}\pi\theta$, then $E_{2n} = 4\displaystyle\int_0^\infty \frac{\theta^{2n}d\theta}{\cosh \frac{1}{2}\pi\theta}$;

and put $\quad u = \pi\phi$, then $B_n = \dfrac{8n}{2^{2n}-1}\displaystyle\int_0^\infty \dfrac{\phi^{2n-1}d\phi}{\sinh \pi\phi}.$

108. *The Remainder in Taylor's Series.*

The previous expansions extend to an infinite number of terms, and are therefore only true when *convergent*.

But some functions, for instance $\sec^{-1}x$, $\cosh^{-1}x$, or $\coth^{-1}x$, cannot be expanded in an infinite series in ascending powers of x, because x must be greater than unity, and the expansion by Taylor's or Maclaurin's Theorem would be *divergent*, and the theorem is then said to fail.

This difficulty will be avoided if we can make the series terminate after a finite number of terms; we shall proceed to explain how this can be done.

Suppose $f(a+h)$ expanded, in Taylor's Series,

$$f(a+h)=fa+hf'a+\frac{h^2}{2!}f''a+\ldots\ldots+\frac{h^n}{n!}f^na+R,$$

where $\qquad R=\frac{h^{n+1}}{(n+1)!}f^{n+1}a+\frac{h^{n+2}}{(n+2)!}f^{n+2}a+\ldots\ldots$

Since all the terms of R are divisible by $\dfrac{h^{n+1}}{(n+1)!}$, we can put $\qquad\qquad R=\dfrac{h^{n+1}}{(n+1)!}P,$

and seek to determine an expression for P.

We shall prove that we can put $P=f^{n+1}(a+\theta h)$, where θ is a proper fraction, some unknown function of a and h; then

$$R=\frac{h^{n+1}}{(n+1)!}f^{n+1}(a+\theta h),$$

and R is called *Lagrange's Form of the Remainder* in Taylor's Series; so that Taylor's Theorem may now be written

$$f(a+h)=fa+hf'a+\frac{h^2}{2!}f''a+\ldots+\frac{h^n}{n!}f^na+\frac{h^{n+1}}{(n+1)!}f^{n+1}(a+\theta h),$$

thus avoiding the use of an infinite series.

Put $a=0$, and change h into x; then

$$fx = f0 + x\,f'0 + \frac{x^2}{2!}f''0 + \ldots + \frac{x^n}{n!}f^n0 + \frac{x^{n+1}}{(n+1)!}f^{n+1}(\theta x),$$

Maclaurin's Theorem with Lagrange's Remainder.

To prove that $\quad P = f^{n+1}(a+\theta h),$

we assume a function Fx, such that

$$Fx = fx + (a+h-x)\,f'x + \frac{(a+h-x)^2}{2!}f''x + \ldots$$

$$+ \frac{(a+h-x)^n}{n!}f^nx + \frac{(a+h-x)^{n+1}}{(n+1)!}P;$$

then $\qquad\qquad F'x = \frac{(a+h-x)^n}{n!}(f^{n+1}x - P);$

also $\qquad Fa = f(a+h)$ and $F(a+h) = f(a+h).$

If we draw the curve BQK, whose equation is $y = Fx$,
then if $\qquad OA = a, \qquad AB = Fa \qquad = f(a+h);$
and if $\qquad OH = a+h, \quad HK = F(a+h) = f(a+h);$
and the chord BK is therefore parallel to the axis of x.

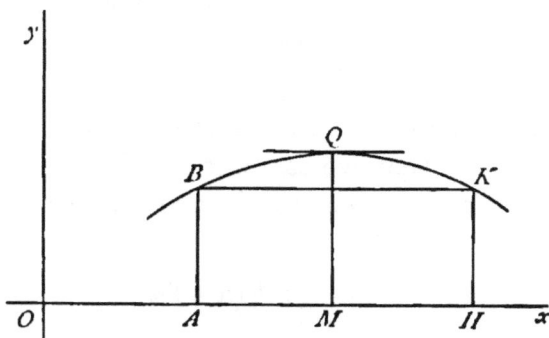

Fig. 37.

Now if fx, $f'x$, $f''x$,... f^nx are all finite and continuous
between $x=a$ and $x=a+h$, then at some point Q of the
curve BQK between B and K the tangent is parallel to
the axis of x.

If $a + \theta h$ is the abscissa of this point Q, θ is a proper fraction, and $AM = \theta h$; then $F'(a + \theta h) = 0$,

or
$$\frac{(h - \theta h)^n}{n!} \{ f^{n+1}(a + \theta h) - P \} = 0,$$

and therefore $P = f^{n+1}(a + \theta h)$.

As an exercise, determine R in the expansion of $(a + x)^m$, $\sin x$, $\cos x$, e^x, a^x, $\sinh x$, $\cosh x$, $\log(1 + x)$, $\tan^{-1} x$, $\tanh^{-1} x$.

We can prove that θ has the limit $\dfrac{1}{n+2}$ when $h = 0$;

for
$$P = f^{n+1} a + \frac{h}{n+2} f^{n+2} a + \ldots = f^{n+1}(a + \theta h)$$
$$= f^{n+1} a + \theta h \, f^{n+2} a + \ldots,$$

so that $\theta = \dfrac{1}{n+2}$, as h becomes indefinitely small.

If we had put $R = hP$, and
$$Fx = fx + (a + h - x)f'x + \frac{(a+h-x)^2}{2!} f''x + \ldots \ldots$$
$$+ \frac{(a+h-x)^n}{n!} f^n x + (a + h - x)P,$$

then
$$F'x = \frac{(a+h-x)^n}{n!} f^{n+1} x - P,$$

so that $F'(a + \theta h) = 0$, when $P = \dfrac{h^n}{n!}(1 - \theta)^n f^{n+1}(a + \theta h)$, and

then
$$R = \frac{h^{n+1}}{n!}(1 - \theta)^n f^{n+1}(a + \theta h),$$

Cauchy's Form of the Remainder in Taylor's Series.

More generally, if we had put $R = h^{p+1} P$, and
$$Fx = fx + (a + h - x)f'x + \frac{(a+h-x)^2}{2!} f''x + \ldots \ldots$$
$$+ \frac{(a+h-x)^n}{n!} f^n x + (a + h - x)^{p+1} P;$$

then $\qquad F'x = \dfrac{(a+h-x)^n}{n!}f^{n+1}x - (p+1)(a+h-x)^p P,$

so that $\qquad\qquad F'(a+\theta h) = 0,$

when $\qquad P = \dfrac{h^{n-p}}{n!(p+1)}(1-\theta)^{n-p}f^{n+1}(a+\theta h),$

and then $\qquad R = \dfrac{h^{n+1}}{n!(p+1)}(1-\theta)^{n-p}f^{n+1}(a+\theta h);$

Schlömilch and Roche's Form of the Remainder.

When $p = n$, this becomes Lagrange's Remainder.

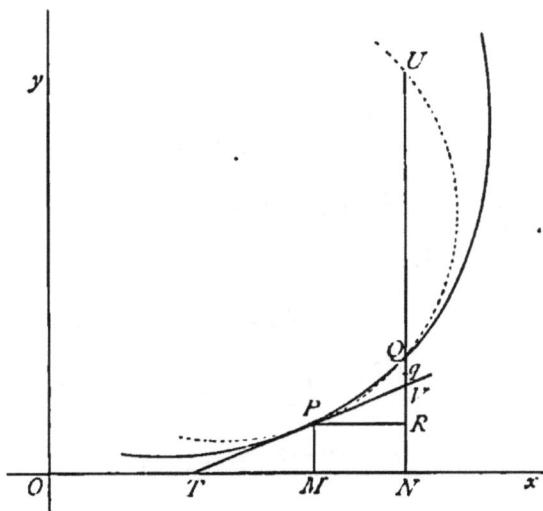

Fig.38.

109. *Geometrical illustration of Taylor's Theorem.*

If $y = fx$ is the equation of a curve PQ, and if $OM = x$, then $MP = fx$ (fig. 38).

If $MN = h$, then $ON = x+h$, and $NQ = f(x+h)$.

Draw the tangent TPV at P, cutting NQ in V, and draw PR parallel to Ox.

Then $\qquad\qquad \tan RPV = f'x,$

so that $RV = hf'x$, and $NV = fx + hf'x$.

But, by Taylor's Theorem,

$$f(x+h) = fx + hf'x + \tfrac{1}{2}h^2 f''(x+\theta h),$$

expanding as far as three terms; so that

$$VQ = \tfrac{1}{2}h^2 f''(x+\theta h).$$

Describe a circle touching the curve at P and passing through Q, and let VQ produced meet the circle in U.

Then, since $\qquad PV^2 = VU \cdot VQ,$

therefore $\qquad VU = \dfrac{PV^2}{VQ} = \dfrac{PR^2 + RV^2}{VQ}$

$$= \frac{h^2\{1+(f'x)^2\}}{\tfrac{1}{2}h^2 f''(x+\theta h)} = 2\frac{1+(f'x)^2}{f''(x+\theta h)}.$$

Now make Q approach to coincidence with P by diminishing h to zero; the circle becomes the circle of curvature at P, and VU becomes the chord of curvature parallel to the axis of y, and equal as before (§ 91).

$$2\frac{1+(f'x)^2}{f''x} = 2\frac{1+\dfrac{dy^2}{dx^2}}{\dfrac{d^2y}{dx^2}}.$$

Again, supposing

$$f(x+h) = fx + hf'x + \frac{h^2}{2!}f''x + \frac{h^3}{3!}f'''(x+\theta h),$$

expanding by Taylor's Theorem as far as four terms; and if Pq is the arc of the parabola, which has its axis parallel to the axis of y, and which *osculates* the curve PQ at P, that is, which has the same circle of curvature at P; then $Vq = \tfrac{1}{2}h^2 f''x$, so that $qQ = \tfrac{1}{6}h^3 f'''(x+\theta h)$.

The geometrical interpretation of

$$f(x+h) = fx + hf'(x+\theta h)$$

has been given in Chapter I., § 10.

110. *Indeterminate Forms.*

In general the value of a function of x is determinate, and is obtained by substitution for any particular value of x.

But when for a particular value of x, say $x = a$, the function assumes one of the *indeterminate forms*,

$$\frac{0}{0}, \ \frac{\infty}{\infty}, \ \infty \times 0, \ \infty - \infty, \ 1^\infty, \ \infty^0, \ 0^0,$$

the real value must be obtained by the *method of limits;* that is, the value of the function must be found for $x = a + h$, where h is small, and reduced and simplified as much as possible by cancelling factors, etc., and afterwards h must be made to vanish.

We have seen this exemplified in Chapter I. in finding the d.c.'s of the simple functions; for

$$\frac{dfx}{dx} = \mathrm{lt} \ \frac{f(x+h) - fx}{h}$$

assumes at first the indeterminate form $\dfrac{0}{0}$, before reduction.

By ordinary algebraical and trigonometrical reductions the indeterminate form may in general be evaluated; but the Differential Calculus affords a general method.

Thus, suppose that when $x = a$, the function $\dfrac{fx}{Fx}$ assumes the indeterminate form $\dfrac{0}{0}$, because $fa = 0$ and $Fa = 0$.

Then, when $x = a$,

$$\frac{\mathrm{f}x}{\mathrm{F}x} = \mathrm{lt}\ \frac{\mathrm{f}(a+h)}{\mathrm{F}(a+h)}$$

$$= \mathrm{lt}\ \frac{\mathrm{f}a + h\mathrm{f}'a + \dfrac{h^2}{2!}\mathrm{f}''a + \dots}{\mathrm{F}a + h\mathrm{F}'a + \dfrac{h^2}{2!}\mathrm{F}''a + \dots}$$

$$= \mathrm{lt}\ \frac{\mathrm{f}'a + \dfrac{h}{2!}\mathrm{f}''a + \dots}{\mathrm{F}'a + \dfrac{h}{2!}\mathrm{F}''a + \dots} = \frac{\mathrm{f}'a}{\mathrm{F}'a}.$$

If however $\mathrm{f}'a = 0$ and $\mathrm{F}'a = 0$, the true value is $\dfrac{\mathrm{f}''a}{\mathrm{F}''a}$; and so on till the value is obtained.

Thus, when $x = a$,

$$\frac{x^n - a^n}{x - a} = \frac{0}{0} = \mathrm{lt}\ \frac{nx^{n-1}}{1} = na^{n-}.$$

Or, putting $x = a + h$,

$$\mathrm{lt}\ \frac{x^n - a^n}{x - a} = \mathrm{lt}\ \frac{(a+h)^n - a^n}{h} = na^{n-1},$$

as in § 3.

If n is a positive integer, then $x - a$ divides $x^n - a^n$, and the quotient,

$$x^{n-1} + ax^{n-2} + a^2x^{n-3} + \dots\dots + a^{n-2}x + a^{n-1},$$

becomes na^{n-1}, when $x = a$.

Again, when $x = 0$,

$$\frac{\sinh x - \sin x}{x^3} = \frac{0}{0},$$

$$= \mathrm{lt}\ \frac{\cosh x - \cos x}{3x^2} = \frac{0}{0},$$

$$= \mathrm{lt}\ \frac{\sinh x + \sin x}{6x} = \frac{0}{0},$$

$$= \mathrm{lt}\ \frac{\cosh x + \cos x}{6} = \frac{1}{3}.$$

Using the expansions of $\sin x$ and $\sinh x$, then this function can be evaluated as follows :—when $x=0$,

$$\frac{\sinh x - \sin x}{x^3}$$

$$= \mathrm{lt}\,\frac{x+\dfrac{x^3}{3!}+\dfrac{x^5}{5!}+\dfrac{x^7}{7!}+\ldots\ldots-x+\dfrac{x^3}{3!}-\dfrac{x^5}{5!}+\dfrac{x^7}{7!}-\ldots\ldots}{x^3}$$

$$= 2\mathrm{lt}\,\frac{\dfrac{x^3}{3!}+\dfrac{x^7}{7!}+\dfrac{x^{11}}{11!}+\ldots\ldots}{x^3}$$

$$= 2\mathrm{lt}\left(\frac{1}{3!}+\frac{x^4}{7!}+\frac{x^8}{11!}+\ldots\ldots\right)=\frac{2}{3!}=\frac{1}{3}.$$

The indeterminate form $\frac{\infty}{\infty}$ can be thrown into the form $\frac{0}{0}$ by interchanging numerator and denominator, and is evaluated in the same manner : for suppose $\frac{\phi x}{\psi x}=\frac{\infty}{\infty}$ when $x=a$, because $\phi a=\infty$, $\psi a=\infty$.

Let $\qquad f x=\dfrac{1}{\psi x}$, $\mathrm{F}x=\dfrac{1}{\phi x}$;

then $\qquad f'x=-\dfrac{\psi'x}{(\psi x)^2}$, $\mathrm{F}'x=-\dfrac{\phi'x}{(\phi x)^2}$;

also $\qquad fa=0$, $\mathrm{F}a=0$.

Then $\qquad\dfrac{\phi a}{\psi a}=\dfrac{\infty}{\infty}=\dfrac{fa}{\mathrm{F}a}=\dfrac{0}{0}$

$$= \mathrm{lt}\,\frac{f(a+h)}{\mathrm{F}(a+h)}=\frac{f'a}{\mathrm{F}'a}=\left(\frac{\phi a}{\psi a}\right)^2\frac{\psi'a}{\phi'a},$$

so that $\qquad\dfrac{\phi a}{\psi a}=\dfrac{\phi'a}{\psi'a}$;

and so on, till the limit is obtained.

The indeterminate form $\infty \times 0$ assumes the form $\dfrac{0}{0}$ by throwing the ∞ into the denominator.

Thus, when $x = \infty$,

$$2^x \sin \frac{a}{2^x} = \infty \times 0 = a \frac{\sin \frac{a}{2^x}}{\frac{a}{2^x}} = \frac{0}{0} = a \ (\S 5).$$

The indeterminate form $\infty - \infty$ can also be made to assume the form $\frac{0}{0}$ by reduction of the terms to a single fraction.

Thus, when $x = \frac{1}{2}\pi$,

$$\sec x - \tan x = \infty - \infty$$
$$= \frac{1 - \sin x}{\cos x} = \frac{0}{0},$$
$$= \frac{1 - \sin x}{\sqrt{(1 - \sin^2 x)}} = \sqrt{\frac{1 - \sin x}{1 + \sin x}} = 0 ;$$

and $\qquad \frac{1}{2}\pi \sec x - x \tan x = \infty - \infty$

$$= \frac{\frac{1}{2}\pi - x \sin x}{\cos x} = \frac{0}{0}$$
$$= \mathrm{lt} \frac{-\sin x - x \cos x}{-\sin x} = 1.$$

To evaluate 1^∞, ∞^0, 0^0, take the logarithm, which will be found to assume the form $\frac{0}{0}$ or $\frac{\infty}{\infty}$, and can then be evaluated by the preceding rules.

For instance, when $x = 0$, $(\cos mx)^{\frac{n}{x^2}} = 1^\infty$; and taking logarithms,

$$\log (\cos mx)^{\frac{n}{x^2}} = n \frac{\log \cos mx}{x^2} = \frac{0}{0}$$
$$= n \, \mathrm{lt} \frac{-m \tan mx}{2x} = \frac{0}{0}$$
$$= n \, \mathrm{lt} \frac{-m^2 \sec^2 mx}{2} = -\tfrac{1}{2} m^2 n,$$

and therefore $\quad (\cos mx)^{\frac{n}{x^2}} = \exp(-\tfrac{1}{2} m^2 n).$

o

Again, when $x = \infty$, $\left(1 - \dfrac{a}{x}\right)^x = 1^\infty$;

and $\log\left(1 - \dfrac{a}{x}\right)^x = x \log\left(1 - \dfrac{a}{x}\right) = \infty \times 0$

$$= \frac{\log\left(1 - \dfrac{a}{x}\right)}{\dfrac{1}{x}} = \frac{0}{0} = -a;$$

so that $\left(1 - \dfrac{a}{x}\right)^x = e^{-a}$.

Otherwise, when $x = \infty$,

$$\text{lt}\left(1 - \frac{a}{x}\right)^x = \text{lt}\left\{\left(1 - \frac{a}{x}\right)^{-\frac{x}{a}}\right\}^{-a} = e^{-a} \ (\S\ 26).$$

Examples.—Prove that, when

(1) $x = 2$, $\dfrac{x^3 - 19x + 30}{x^3 - 2x^2 - 9x + 18} = \dfrac{7}{5}$;

     $x = 3$, ... $= \dfrac{4}{3}$.

(2) $x = a$, $\dfrac{\sqrt{a} - \sqrt{x} + \sqrt{(a-x)}}{\sqrt{(a^2 - x^2)}} = \dfrac{1}{\sqrt{(2a)}}$.

(3) $x = 0$, $\dfrac{\tan x + \sec x - 1}{\tan x - \sec x + 1} = 1$.

(4) $x = \frac{1}{2}\pi$, $\dfrac{\cos x + 1 - \sin x}{\cos x - 1 + \sin x} = 1$.

(5) $x = \frac{1}{4}\pi$, $\dfrac{\sin x - \cos x}{\sin 2x - \cos 2x - 1} = \frac{1}{2}\sqrt{2}$.

(6) $x = \frac{1}{6}\pi$, $\dfrac{1 - \sin x - 2\sin^2 x}{1 - 3\sin x + 2\sin^2 x} = 3$.

(7) $x = 1$, $\dfrac{\log x}{x - 1} = 1$.

(8) $x = 0$, $\dfrac{\sinh x}{\log(1 + x)} = 1$.

(9) $x = 0$, $\dfrac{1 - \cos x}{x^2} = \frac{1}{2}$.

(10) $x = 0$, $\dfrac{(1+x)^{\frac{1}{x}} - e + \frac{1}{2}ex - \frac{11}{24}ex^2}{x^3} = -\frac{11}{24}e$.

(11) $x = 0$, $\dfrac{\tan(\sin x) - \sin(\tan x)}{x^7} = \dfrac{1}{30}$.

(12) $x = a$,

$$\dfrac{fx - fa}{Fx - Fa} = \dfrac{f'a}{F'a};$$

$$\dfrac{fx - fa - (x - a) f'a}{Fx - Fa - (x - a) F'a} = \dfrac{f''a}{F''a};$$

$$\dfrac{fx - fa - (x - a) f'a - \frac{1}{2}(x - a)^2 f''a}{Fx - Fa - (x - a) F'a - \frac{1}{2}(x - a)^2 F''a} = \dfrac{f'''a}{F'''a};$$

and so on.

(13) $x = \infty$, $\dfrac{ax^m + bx^{m-1} + cx^{m-2} + \ldots}{Ax^n + Bx^{n-1} + Cx^{n-2} + \ldots} = 0$, $\dfrac{a}{A}$, or ∞,

according as m is $<, =$, or $> n$.

(14) $x = \frac{1}{2}\pi$, $\dfrac{\sec(2n+1)x}{\sec(2m+1)x} = (-1)^{m-n}\dfrac{2m+1}{2n+1}$,

if m and n are integers.

(15) $x = 0$, $\dfrac{\log \cot x}{\operatorname{cosec} x} = 0$.

(16) $x = 0$, $x^n \log x = 0$, if n is positive, $= -\infty$, if n is negative.

(17) $x = 1$, $(1 - x)\tan \frac{1}{2}\pi x = \dfrac{2}{\pi}$.

(18) $x = \infty$, $x(a^{\frac{1}{x}} - 1) = \log a$.

(19) $x = 1$, $\dfrac{x}{\log x} - \dfrac{1}{\log x} = 1$.

(20) $x = 0$, $\operatorname{cosec}^2 x - \dfrac{1}{x^2} = \frac{1}{3}$.

(21) $x = 0$, $(\cos ax)^{\operatorname{cosec}^2 bx} = \exp\left(-\tfrac{1}{2}\frac{a^2}{b^2}\right)$.

(22) $x = 0$, $\left(\dfrac{\sin x}{x}\right)^{\frac{1}{x}} = 1$, $\left(\dfrac{\sin x}{x}\right)^{\frac{1}{x^2}} = e^{-\frac{1}{6}}$, $\left(\dfrac{\sin x}{x}\right)^{\frac{1}{x^3}} = 0$.

(23) $x = \tfrac{1}{2}\pi$, $(\sin x)^{\tan x} = 1$.

(24) $x = 0$, $(\cot x)^{\sin x} = 1$.

(25) $x = 0$, $x^x = 1$.

(26) $x = 0$, $(\sin x)^{\tan x} = 1$.

111. *Fourier's Series.*

In many physical problems it is requisite to expand a function $f x$, not in ascending powers of x as in Taylor's series, but in a series of cosines and sines of multiples of x, between certain limits.

Suppose then, that, between the limits $x = \pm l$, $f x$ can be expanded in the form

$$f x = \tfrac{1}{2}A_0 + A_1 \cos \pi\frac{x}{l} + A_2 \cos 2\pi\frac{x}{l} + \ldots + A_n \cos n\pi\frac{x}{l} + \ldots\ldots$$

$$+ B_1 \sin \pi\frac{x}{l} + B_2 \sin 2\pi\frac{x}{l} + \ldots\ldots + B_n \sin n\pi\frac{x}{l} + \ldots\ldots,$$

it is required to determine the value of A_n and B_n.

Since the cosine is an *even* function of x, and the sine an *odd* function of x (§ 50), therefore dividing $f x$ into its even part $\tfrac{1}{2}\{f(x) + f(-x)\}$ and its odd part $\tfrac{1}{2}\{f(x) - f(-x)\}$, it follows that

$$\tfrac{1}{2}\{f(x) + f(-x)\}$$

$$= \tfrac{1}{2}A_0 + A_1 \cos \pi\frac{x}{l} + A_2 \cos 2\pi\frac{x}{l} + \ldots + A_n \cos n\pi\frac{x}{l} + \ldots \text{(i.)};$$

$$\tfrac{1}{2}\{f(x) - f(-x)\}$$

$$= \qquad B_1 \sin \pi\frac{x}{l} + B_2 \sin 2\pi\frac{x}{l} + \ldots + B_n \sin n\pi\frac{x}{l} + \ldots \text{(ii.)}.$$

Now, if m and n are unequal integers, then (§ 42)

$$\int \cos m\theta \cos n\theta d\theta = \frac{1}{2} \frac{\sin (m-n)\theta}{m-n} + \frac{1}{2} \frac{\sin (m+n)\theta}{m+n},$$

$$\int \sin m\theta \sin n\theta d\theta = \frac{1}{2} \frac{\sin (m-n)\theta}{m-n} - \frac{1}{2} \frac{\sin (m+n)\theta}{m+n},$$

so that

$$\int_0^\pi \cos m\theta \cos n\theta = 0, \quad \int_0^\pi \sin m\theta \sin n\theta d\theta = 0.$$

But, if $m=n$,

$$\int_0^\pi \cos^2 n\theta d\theta = \tfrac{1}{2}\pi, \quad \int_0^\pi \sin^2 n\theta d\theta = \tfrac{1}{2}\pi.$$

Similarly, if m and n are unequal integers,

$$\int_0^l \cos m\pi\frac{v}{l} \cos n\pi\frac{v}{l}dv = 0, \quad \int_0^\pi \sin m\pi\frac{v}{l} \sin n\pi\frac{v}{l}dv = 0 ;$$

but, if $m=n$,

$$\int_0^l \cos^2 n\pi\frac{v}{l}dv = \tfrac{1}{2}l, \quad \int_0^l \sin^2 n\pi\frac{v}{l}dv = \tfrac{1}{2}l;$$

also

$$\int_0^l \sin n\pi\frac{v}{l} \cos n\pi\frac{v}{l}dv = 0.$$

Therefore, to determine A_n, multiply both sides of the corresponding equation (i.) by $\cos n\pi\frac{x}{l}$, and integrate between the limits 0 and l; then, changing x into v under the sign of integration,

$$A_n = \frac{1}{l} \int_0^l \{f(v) + f(-v)\} \cos n\pi\frac{v}{l}dv,$$

and

$$A_0 = \frac{1}{l} \int_0^l \{f(v) + f(-v)\}dv.$$

Similarly from equation (ii.),

$$B_n = \frac{1}{l} \int_0^l \{f(v) - f(-v)\}\sin n\pi\frac{v}{l}dv.$$

Therefore, between the limits $x = \pm l$,

$$fx = \frac{1}{l}\int_0^l \tfrac{1}{2}\{f(v) + f(-v)\}\,dv$$

$$+ \frac{1}{l}\sum_{n=1}^{n=\infty}\left[\int_0^l \{f(v) + f(-v)\}\cos n\pi\frac{v}{l}dv\right]\cos n\pi\frac{x}{l}$$

$$+ \frac{1}{l}\sum_{n=1}^{n=\infty}\left[\int_0^l \{f(v) - f(-v)\}\sin n\pi\frac{v}{l}dv\right]\sin n\pi\frac{x}{l};$$

which is called *Fourier's Series.*

At the limits $x = \pm l$, the value of the series is $\tfrac{1}{2}\{f(l) + f(-l)\}$, so that there is discontinuity in general at the limits; and outside these limits Fourier's series represents periodic repetitions of the function fx between the limits.

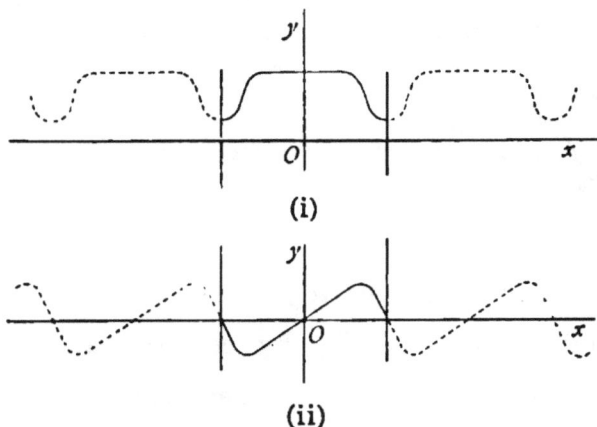

(i)

(ii)

Fig. 39.

This is exhibited by drawing the curves representing the series

$$\tfrac{1}{2}A_0 + A_1\cos\pi\frac{x}{l} + A_2\cos 2\pi\frac{x}{l} + \ldots\ldots A_n\cos n\pi\frac{x}{l} + \ldots\ldots \text{(i.)}$$

and $$B_1\sin\pi\frac{x}{l} + B_2\sin 2\pi\frac{x}{l} + \ldots\ldots B_n\sin n\pi\frac{x}{l} + \ldots\ldots\text{(ii.)}$$

in fig. 39, (i.) and (ii.) respectively.

When the limits between which the function fx is to be expanded are 0 and l, then either series (i.) proceeding in cosines of multiples of x, or series (ii.) proceeding in sines of multiples of x, may be employed at pleasure, but it is best to choose the series which introduces the least discontinuity at the limits.

For instance, suppose $fx = x$; then between 0 and l

$$x = l - \frac{l}{(\frac{1}{2}\pi)^2}\left(\cos \pi\frac{x}{l} + \frac{1}{3^2}\cos 3\pi\frac{x}{l} + \frac{1}{5^2}\cos 5\pi\frac{x}{l} + \ldots\right),$$

or $\quad x = \dfrac{l}{\frac{1}{2}\pi}\left(\sin \pi\dfrac{x}{l} - \dfrac{1}{2}\sin 2\pi\dfrac{x}{l} + \dfrac{1}{3}\sin 3\pi\dfrac{x}{l} - \ldots\right);$

but outside these limits the series represent the curves drawn in fig. 40, (i.) and (ii.).

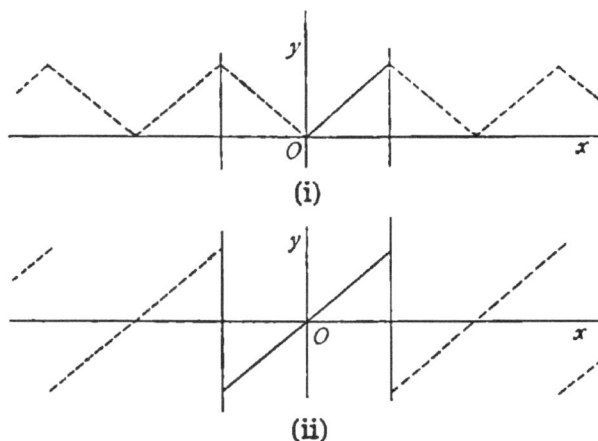

(i)

(ii)

Fig.40.

The student is recommended to trace the curves represented by three or four of the leading terms of these expansions, in order to see how quickly the series approximates to the given function of which it is the expansion.

Examples.—(1) Expand x^2 in a series of cosines of multiples of $\pi\frac{x}{l}$; then

$$x^2 = \tfrac{1}{4}l^2 - \frac{l^2}{(\tfrac{1}{2}\pi)^3}\left(\cos \pi\frac{x}{l} - \frac{1}{3^3}\cos 3\pi\frac{x}{l} + \frac{1}{5^3}\cos 5\pi\frac{x}{l} - \ldots\right).$$

(2) Expand $\cosh mx$ in a series of cosines; then

$\cosh mx$

$$= 2\pi\sinh ml\left(\frac{1}{2m^2l^2} - \frac{\cos \pi\frac{x}{l}}{m^2l^2 + \pi^2} + \frac{\cos 2\pi\frac{x}{l}}{m^2l^2 + 2^2\pi^2} - \frac{\cos 3\pi\frac{x}{l}}{m^2l^2 + 3^2\pi^2} + \ldots\right).$$

(3) Expand $\sinh mx$ in a series of sines; then

$\sinh mx$

$$= 2\pi\sinh ml\left(\frac{\sin \pi\frac{x}{l}}{m^2l^2 + \pi^2} - \frac{2\sin 2\pi\frac{x}{l}}{m^2l^2 + 2^2\pi^2} + \frac{3\sin 3\pi\frac{x}{l}}{m^2l^2 + 3^2\pi^2} - \ldots\right).$$

This expansion of $\sinh mx$ can be derived by differentiation from the preceding expansion of $\cosh mx$.

(4) Expand $\cos mx$ in a series of cosines, and $\sin mx$ in a series of sines.

Examine the case when ml is a multiple of π.

CHAPTER V.

PARTIAL DIFFERENTIATION.

112. *Functions of Two Independent Variables.*

When y is a function fx of a single variable x, the relation $y = fx$ is exhibited by means of a plane curve in which the abscissa is x, and corresponding ordinate $y = fx$ (§ 9).

But when a variable quantity z is a function of two independent variables x and y, expressed by the notation

$$z = f(x, y),$$

then x and y may be supposed to be the co-ordinates of any point on a datum (horizontal) plane, and z to be the height of a surface above this plane; and then

$$z = f(x, y)$$

is the equation of the surface.

Now suppose a vertical section of the surface made by a vertical plane parallel to the axis of x; then the tangent of the slope of the curve of section to the horizon will be the d.c. of z with respect to x, keeping y constant, and is expressed by

$$\frac{\partial z}{\partial x} \text{ or } \frac{\partial f(x, y)}{\partial x} \left(\text{generally abbreviated to } \frac{\partial f}{\partial x} \right),$$

and this is called the *partial* d.c. of z with respect to x, ∂ being used when the differentiation is *partial*.

Again, if a vertical section of the surface is made by a vertical plane parallel to the axis of y, then the tangent of the slope of the curve of section to the horizon will be the d. c. of z with respect to y, keeping x constant, and is expressed by

$$\frac{\partial z}{\partial y} \text{ or } \frac{\partial f(x, y)}{\partial y} \left(\text{abbreviated to } \frac{\partial f}{\partial y}\right).$$

Now, to find $\frac{dz}{dt}$, the rate of increase of z, when x and y increase at given rates $\frac{dx}{dt}$ and $\frac{dy}{dt}$, suppose Δx and Δy to be small finite increments of x and y, and Δz the corresponding increment of z.

Then

$$z + \Delta z = f(x + \Delta x, y + \Delta y),$$
$$\Delta z = f(x + \Delta x, y + \Delta y) - f(x, y),$$
$$\text{and } \frac{\Delta z}{\Delta t} = \frac{f(x + \Delta x, y + \Delta y) - f(x, y)}{\Delta t}$$
$$= \frac{f(x + \Delta x, y + \Delta y) - f(x, y + \Delta y) + f(x, y + \Delta y) - f(x, y)}{\Delta t}$$
$$= \frac{f(x + \Delta x, y + \Delta y) - f(x, y + \Delta y)}{\Delta x} \frac{\Delta x}{\Delta t}$$
$$+ \frac{f(x, y + \Delta y) - f(x, y)}{\Delta y} \frac{\Delta y}{\Delta t}.$$

Proceeding to the limit, then

$$\text{lt} \frac{f(x + \Delta x, y + \Delta y) - f(x, y + \Delta y)}{\Delta x}$$
$$= \text{lt} \frac{\partial f(x, y + \Delta y)}{\partial x} = \frac{\partial f(x, y)}{\partial x} = \frac{\partial z}{\partial x};$$

and $$\operatorname{lt}\frac{f(x,\,y+\Delta y)-f(x,\,y)}{\Delta y}=\frac{\partial f(x,\,y)}{\partial y}=\frac{\partial z}{\partial y}\,;$$

therefore $$\frac{dz}{dt}=\frac{\partial z}{\partial x}\frac{dx}{dt}+\frac{\partial z}{\partial y}\frac{dy}{dt}\,.$$

Differentiating again, according to this rule,

$$\frac{d^2z}{dt^2}=\frac{\partial z}{\partial x}\frac{d^2x}{dt^2}+\frac{\partial z}{\partial y}\frac{d^2y}{dt^2}$$

$$+\frac{\partial^2 z}{\partial x^2}\frac{dx^2}{dt^2}+2\frac{\partial^2 z}{\partial x\partial y}\frac{dx}{dt}\frac{dy}{dt}+\frac{\partial^2 z}{\partial y^2}\frac{dy^2}{dt^2}\,;$$

and so on.

Suppose $z=c$, a constant; then

$$f(x,y)=c$$

is the *implicit relation* (§ 15) connecting x and y along the curve of section of the surface $z=f(x,y)$ by the plane $z=c$, and is therefore the equation of the curve.

Then, along this curve, $\dfrac{dz}{dt}=0$, and therefore

$$\frac{\partial f}{\partial x}\frac{dx}{dt}+\frac{\partial f}{\partial y}\frac{dy}{dt}=0\,;$$

or with x instead of t for the independent variable

$$\frac{\partial f}{\partial x}+\frac{\partial f}{\partial y}\frac{dy}{dx}=0,$$

the *first derived equation* (§ 15).

This can be proved independently; for if Δx and Δy denote simultaneous increments of x and y, then

$$f(x+\Delta x,\ y+\Delta y)=c,$$

or $$f(x+\Delta x,\ y+\Delta y)-f(x,y)=0\,;$$

which can be written

$$\frac{f(x+\Delta x,\ y+\Delta y)-f(x,\ y+\Delta y)}{\Delta x}+\frac{f(x,\ y+\Delta y)-f(x,\ y)}{\Delta y}\frac{\Delta y}{\Delta x}=0\,;$$

and proceeding to the limit by making Δx and Δy ultimately zero,

$$\frac{\partial f}{\partial x} + \frac{\partial f}{\partial y}\frac{dy}{dx} = 0,$$

as before.

The *second derived equation* is found by differentiating again with respect to x by this rule, and is therefore

$$\frac{\partial^2 f}{\partial x^2} + 2\frac{\partial^2 f}{\partial x \partial y}\frac{dy}{dx} + \frac{\partial^2 f}{\partial y^2}\frac{dy^2}{dx^2} + \frac{\partial f}{\partial y}\frac{d^2 y}{dx^2} = 0;$$

and so on, for the third, fourth, ... derived equations.

113. *Expansion of a Function of two Independent Variables.*

Let $z = f(x, y)$, and let x receive an increment h, and y an increment k, to find z_1, the new value of z.

Then $z_1 = f(x+h, y+k)$, and expanding by Taylor's Theorem, first in powers of h,

$$z_1 = f(x, y+k) + h\frac{\partial}{\partial x}f(x, y+k) + \frac{h^2}{2!}\frac{\partial^2}{\partial x^2}f(x, y+k) + \dots$$

and now expanding each term by Taylor's Theorem in powers of k, and writing the terms in each series diagonally,

$$z_1 = f(x, y) + h\frac{\partial f}{\partial x} + \frac{h^2}{2!}\frac{\partial^2 f}{\partial x^2} + \dots$$
$$+ k\frac{\partial f}{\partial y} + hk\frac{\partial^2 f}{\partial x \partial y} + \dots$$
$$+ \frac{k^2}{2!}\frac{\partial^2 f}{\partial y^2} + \dots$$

and the general term will be found to be

$$\frac{1}{n!}\left(h^n\frac{\partial^n f}{\partial x^n} + nh^{n-1}k\frac{\partial^n f}{\partial x^{n-1}\partial y} + \dots + k^n\frac{\partial^n f}{\partial y^n}\right)$$

which may be written in the abbreviated symbolical form

$$\frac{1}{n!}\left(h\frac{\partial}{\partial x} + k\frac{\partial}{\partial y}\right)^{n} f(x, y).$$

Since it is immaterial whether we expand first with respect to x and then with respect to y, or in the reverse order, it is seen that the order of partial differentiation is immaterial, or

$$\frac{\partial^2 z}{\partial x\,\partial y} = \frac{\partial^2 z}{\partial y\,\partial x}.$$

This may be proved independently from the definition ;

for $\qquad \dfrac{\partial z}{\partial y} = \mathrm{lt}\dfrac{f(x, y+k) - f(x, y)}{k}$,

and $\quad \dfrac{\partial^2 z}{\partial x\,\partial y} = \mathrm{lt}\dfrac{f(x+h,\, y+k) - f(x+h,\, y) - f(x, y+k) + f(x, y)}{hk}$;

and $\dfrac{\partial^2 z}{\partial y\,\partial x}$ is the limit of the same expression.

The general form of the expansion can be more easily perceived by putting

$$Ft = f(x + ht,\, y + kt)$$

and expanding in powers of t by Maclaurin's Theorem, and then putting $t = 1$.

Now by Maclaurin's Theorem

$$Ft = F0 + tF'0 + \frac{t^2}{2!}F''0 + \dots + \frac{t^n}{n!}F^n0 + \frac{t^{n+1}}{(n+1)!}F^{n+1}(\theta t) ;$$

and $\qquad F't = h\dfrac{\partial Ft}{\partial x} + k\dfrac{\partial Ft}{\partial y}$

$$F''t = h^2\frac{\partial^2 Ft}{\partial x^2} + 2hk\frac{\partial^2 Ft}{\partial x\,\partial y} + k^2\frac{\partial^2 Ft}{\partial y^2},$$

which may be written symbolically (§ 78)

$$F''t = \left(h\frac{\partial}{\partial x} + k\frac{\partial}{\partial y}\right)^2 Ft ;$$

and generally

$$F^n t = \left(h \frac{\partial}{\partial x} + k \frac{\partial}{\partial y} \right)^n Ft;$$

so that, putting $t = 1$,

$$f(x + h, y + k) = f(x, y) + \left(h \frac{\partial}{\partial x} + k \frac{\partial}{\partial y} \right) f(x, y) + \cdots$$

$$+ \frac{1}{n!} \left(h \frac{\partial}{\partial x} + k \frac{\partial}{\partial y} \right)^n f(x, y) + \frac{1}{(n+1)!} \left(h \frac{\partial}{\partial x} + k \frac{\partial}{\partial y} \right)^{n+1} f(x + \theta h, y + \theta k).$$

114. *Maxima and Minima of a Function of two Independent Variables.*

Suppose the equation $z = f(x, y)$ to represent the surface of a country, and to be cut into a series of contour lines by a system of horizontal planes.

Then at the summit of a hill, where z has a maximum value, the contour line shrinks into a point, and

$$\frac{\partial z}{\partial x} = 0, \quad \frac{\partial z}{\partial y} = 0.$$

The contour line a little lower down is a small closed curve, and the corresponding horizontal plane cuts off a small cap from the surface.

At the bottom of a lake, where z has a minimum value, the contour line shrinks into a point, and again

$$\frac{\partial z}{\partial x} = 0, \quad \frac{\partial z}{\partial y} = 0.$$

The contour line next above is a small closed curve, and the corresponding horizontal plane cuts off a small cup from the surface.

Starting from the top of a hill and going down, the contour lines enlarge, till a *pass* is reached, where two hills meet; here the contour lines cross, and

$$\frac{\partial z}{\partial x} = 0, \quad \frac{\partial z}{\partial y} = 0;$$

but z is not a maximum or minimum; for z is a minimum with respect to the two adjacent hills, a maximum with respect to the two adjacent valleys.

Again, starting from the bottom of a lake, and going up, the contour lines enlarge till a *bar* is reached, where two depressed regions meet, and the contour lines cross,

and again $$\frac{\partial z}{\partial x} = 0, \ \frac{\partial z}{\partial y} = 0;$$

but z is neither a maximum or minimum.

If there are p tops of hills and q bottoms of lakes, there must be $p-1$ passes, and $q-1$ bars; also there must be at least two summits of hills higher than any pass, and two bottoms of lakes lower than any bar; this is seen by drawing the contour lines. (Maxwell, *Math. Tripos,* 1870.)

These geometrical considerations are useful in the problem of finding the maximum and minimum values of a function of two independent variables, $z = f(x, y)$.

(Rev. E. Hill, *Messenger of Mathematics,* vol. V., p. 84.)

115. *Double Integration.*

Let V denote the volume of the solid which is bounded by the surface $z = f(x, y)$ and a cylinder standing on any base A in the plane of $z = 0$, with generating lines parallel to the axis of z.

Then the *fluent* V may be supposed generated by the motion of the *fluxion* $\frac{\partial V}{\partial y}$, an area moving with its plane perpendicular to the axis of y; and then $\frac{\partial V}{\partial y}$ considered as a fluent may be supposed generated by the motion of the ordinate z, moving parallel to the axis of x, so that z

is the fluxion of $\dfrac{\partial V}{\partial y}$ with respect to x, or

$$\frac{\partial^2 V}{\partial x \partial y} = z \, ;$$

and integrating, $\qquad V = \iint z\, dx\, dy,$

the integration extending over the area of the base A.

In this manner the volume V may be considered as built up of infinitesimal filaments of height z and base $dx\, dy$.

Also, if $\bar{x}$, $\bar{y}$, $\bar{z}$ denote the co-ordinates of the centroid of the volume V,

$$\bar{x} = \frac{\iint xz\, dx\, dy}{\iint z\, dx\, dy}, \quad \bar{y} = \frac{\iint yz\, dx\, dy}{\iint z\, dx\, dy}, \quad \bar{z} = \frac{\iint \frac{1}{2}z^2\, dx\, dy}{\iint z\, dx\, dy}.$$

Applying this to find the volume of the hemisphere bounded by $x^2 + y^2 + z^2 = a^2$ and the plane $z = 0$, and integrating first with respect to x, the limits are $\pm \sqrt{(a^2 - y^2)}$, and integrating afterwards with respect to y, the limits are $\pm a$; so that

$$V = \int_{-a}^{a} \int_{-\sqrt{(a^2-y^2)}}^{\sqrt{(a^2-y^2)}} \sqrt{(a^2 - x^2 - y^2)}\, dx\, dy$$

$$= \int_{-a}^{a} \left\{ \tfrac{1}{2}x\sqrt{(a^2 - x^2 - y^2)} + \tfrac{1}{2}(a^2 - y^2)\sin^{-1}\frac{x}{\sqrt{(a^2 - y^2)}} \right\}_{-\sqrt{(a^2-y^2)}}^{\sqrt{(a^2-y^2)}} dy$$

$$= \tfrac{1}{2}\pi \int_{-a}^{a} (a^2 - y^2)\, dy = \tfrac{1}{2}\pi (a^2 y - \tfrac{1}{3}y^3)_{-a}^{a} = \tfrac{2}{3}\pi a^3.$$

The preceding notation is useful in representing an indefinite integral before it is taken between the given limits.

We may suppose z to denote the variable density of a plane lamina, and then the preceding formulæ will give V, the mass of the lamina, and $\bar{x}$, $\bar{y}$ the co-ordinates of its centre of mass.

If $z = c$, a constant, then V is the volume of a right cylinder on the base A, so that

$$V = c \iint dx dy, \text{ and } A = \iint dx dy.$$

The area A may now be considered as built up of the infinitesimal elements of area $dx dy$, and the limits of integration must be taken so as to include all these elements of area in A.

116. *The Sign to be attributed to an Area.*

Suppose a point to travel once round the closed area A, so as always to have the interior of the curve on the *left* hand (fig. 41).

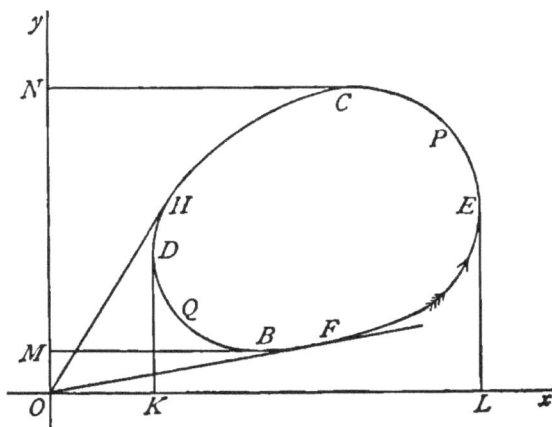

Fig.41.

Then
$$A = \iint dx dy = \int x dy = \int x \frac{dy}{dt} dt,$$

taken round the curve.

From B to C along BPC, $\dfrac{dy}{dt}$ is positive, and

$$\int x \frac{dy}{dt} dt = \text{area } MBPCN$$

P

and from C to B along CQB, $\dfrac{dy}{dt}$ is negative, and

$$\int x\frac{dy}{dt}dt = -\text{area } MBQCN;$$

so that, taken round the curve,

$$\int x\frac{dy}{dt}dt \text{ or } \int x\,dy = A, \text{ the area of the closed curve.}$$

But
$$\iint dy\,dx = \int y\,dx = \int y\frac{dx}{dt}dt;$$

and from E to D along EPD, $\dfrac{dx}{dt}$ is negative, so that

$$\int y\frac{dx}{dt}dt = -\text{area } LEPDK;$$

and from D to E along DQE, $\dfrac{dx}{dt}$ is positive, so that

$$\int y\frac{dx}{dt}dt = \text{area } LEQDK;$$

and therefore, taken round the curve,

$$\int y\frac{dx}{dt}dt \text{ or } \int y\,dx = -A.$$

Therefore, taken round the curve,

$$\int\left(x\frac{dy}{dt} + y\frac{dx}{dt}\right)dt = 0,$$

and
$$\left(x\frac{dy}{dt} + y\frac{dx}{dt}\right)dt = d(xy)$$

is called a *perfect_differential*, and its integral between limits is independent of the intermediate values of x and y, so that taken round any closed curve the integral is zero.

But
$$A = \tfrac{1}{2}\int\left(x\frac{dy}{dt} - y\frac{dx}{dt}\right)dt$$
$$= \tfrac{1}{2}\int(x\,dy - y\,dx).$$

Changing to polar co-ordinates,
$$x = r \cos \theta, \; y = r \sin \theta,$$
then (§ 85)
$$x \frac{dy}{dt} - y \frac{dx}{dt} = r^2 \frac{d\theta}{dt} \; ;$$
so that
$$A = \tfrac{1}{2} \int r^2 \frac{d\theta}{dt} dt = \tfrac{1}{2} \int r^2 d\theta,$$
taken round the curve.

Draw the tangents OF, OH to the curve from O, supposing the origin O is outside the area; then along FPH, $\frac{d\theta}{dt}$ is positive, and
$$\tfrac{1}{2} \int r^2 \frac{d\theta}{dt} dt = \text{area } OFPCHO ;$$
but along HQF, $\frac{d\theta}{dt}$ is negative, and
$$\tfrac{1}{2} \int r^2 \frac{d\theta}{dt} dt = -\text{area } OFQHO,$$
so that, taken round the curve,
$$\tfrac{1}{2} \int r^2 \frac{d\theta}{dt} dt = \tfrac{1}{2} \int r^2 d\theta = A.$$

If the perimeter of the curve cuts itself, then the area obtained by integrating once round a loop will be positive or negative with the above formulæ, according as the area is on the left or right of the point as it travels round the curve.

(Clifford, *Common Sense of the Exact Sciences*).

With polar co-ordinates and double integration
$$A = \iint r \, dr \, d\theta, \; V = \iint z r \, dr \, d\theta;$$
and the area A must be supposed divided into infinitesimal elements of area $r \, dr \, d\theta$ by concentric circles with common centre at O, and straight lines radiating from O.

For if r, θ are the polar co-ordinates of a point, then the finite element of area inclosing the point, bounded by circles of radii $r-\frac{1}{2}\Delta r$ and $r+\frac{1}{2}\Delta r$, and vectors from O making angles $\theta-\frac{1}{2}\Delta\theta$ and $\theta+\frac{1}{2}\Delta\theta$ with the axis of x, is

$$\tfrac{1}{2}\{(r+\tfrac{1}{2}\Delta r)^2-(r-\tfrac{1}{2}\Delta r)^2\}\Delta\theta=r\Delta r\Delta\theta,$$

which ultimately becomes $r\,dr\,d\theta$.

117. *The Planimeter (Amsler's).*

This instrument in the most usual form, that invented by Amsler of Schaffhausen, consists of two bars OA, AP, jointed at A, and carrying in PA produced a small graduated roller R, with axis fixed parallel to PA (fig. 42).

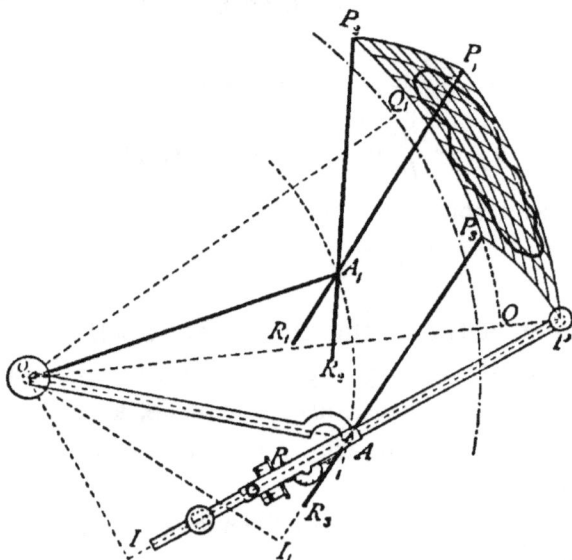

Fig. 42.

The instrument is used to measure areas; to do this O is pivoted at a fixed point, and a pointer at P is carried round the perimeter of the curve whose area is required;

the roller R, which rolls and slides on the plane of the paper, then registers the area.

If precision is required the point P is carried, say ten times round the perimeter, and the reading of the roller divided by ten; but in the so-called *precision* plani-meters the roller always moves on a sheet of prepared cardboard fixed to the instrument, so as to avoid the roughness and irregularity of the surface of ordinary paper, which slightly impair the precision of the instru-ment described here.

To explain the theory of the instrument, let $OA = a$, $AP = b$, $AR = c$, and the radius of the roller $= r$; and let the direction of a positive rotation of the roller, as marked by the graduations, be that of rotation on a right-handed screw on the axle of R which would give a motion in the direction AR.

Drop the perpendicular OI from O on AR, and first suppose the joint A clamped.

Then if I is in AR produced, a rotation of the instrument about O with *angular* velocity $\frac{d\theta}{dt}$ will give to R the com-ponent velocities $OI\frac{d\theta}{dt}$ in the direction IR and $IR\frac{d\theta}{dt}$ perpendicular to IR, and therefore compel the roller to turn with angular velocity $\frac{RI}{r}\frac{d\theta}{dt}$; but when I is on the other side of R, the angular velocity of the roller will be $-\frac{RI}{r}\frac{d\theta}{dt}$.

Therefore, keeping A clamped, the roller will turn through an angle $\frac{RI}{r}\theta$ or $-\frac{RI}{r}\theta$, according as I is not or is on the same side of R as A, when the instrument is rotated through an angle θ about O.

When I coincides with R, the roller will not turn, and then P describes a circle called the zero circle, represented by the middle dotted circular line (fig. 42) of radius

$$\sqrt{(OR^2+RP^2)} = \sqrt{\{a^2-c^2+(b+c)^2\}} = \sqrt{(a^2+b^2+2bc)}.$$

Next unclamp the joint A, and clamp O; the roller will turn with angular velocity $-\dfrac{c}{r}\dfrac{d\phi}{dt}$ when the bar AP turns with angular velocity $\dfrac{d\phi}{dt}$, so that the roller will turn through an angle $-\dfrac{c}{r}\phi$ while AP turns through an angle ϕ.

Now suppose P to travel round the finite circuit $PP_1P_2P_3$ by a combination of the preceding motions in the following order:—

(1) Clamp the joint A, and move P to P_1, and A to A_1 on arcs of circles of centre O; then the roller will turn through an angle $\dfrac{RI}{r}\theta$, if the angle $AOA_1=POP_1=\theta$.

(2) Unclamp A and clamp O, and move P_1 to P_2 on the arc of a circle of centre A_1; then the roller will move through an angle $-\dfrac{c}{r}\phi$, if the angle $P_1A_1P_2=\phi$.

(3) Unclamp O and clamp A, and move P_2 backwards to P_3 and A_1 to A on arcs circles of centre O, through an angle θ; then the roller will move through an angle $-\dfrac{RI_1}{r}\theta$, if OI_1 is the perpendicular from O on P_2A.

(4) Unclamp A and clamp O, and move P_3 to P on the arc of a circle of centre A, and consequently through an angle ϕ; the roller will turn through an angle $\frac{c}{r}\phi$, which cancels the angle due to motion (2).

In completing the finite circuit $PP_1P_2P_3$ the roller will then have turned through an angle

$$(RI - RI_1)\frac{\theta}{r}.$$

But the area $PP_1P_2P_3 =$ area PP_1Q_1Q
$\quad =$ sector $OPP_1 -$ sector OQQ_1
$\quad = \frac{1}{2}(OP^2 - OP_3^2)\theta$
$\quad = \frac{1}{2}(OA^2 + AP^2 + 2AI \cdot AP - OA^2 - AP^2 - 2AI_1 \cdot AP)\theta$
$\quad = (AI - AI_1)b\theta$
$\quad = br$ times the angle (in c.m.) turned through by the roller.

The area $PP_1P_2P_3$ is therefore b times the travel of the circumference of the roller, so that by altering the length b by an adjustment on the instrument, the area can be read off in any required units.

Any irregular area must be supposed to be built up of infinitesimal elements found in the same manner as $PP_1P_2P_3$; and will be accurately measured by the roller when the point P completes a circuit of the perimeter, both joints being now free to turn simultaneously.

If, however, the origin O is insidet he area, the area of the zero circle must be added to the reading of the roller.

A complete account of Planimeters will be found in a paper by Hele Shaw, Professor of Engineering at University College, Liverpool, read before the Institution of Civil Engineers, April, 1885.

118. *Functions of three or more Independent Variables.*

A function of three independent variables, x, y, z, denoted by f(x, y, z), may be supposed to represent some function of the position of a point in space whose co-ordinates are x, y, z; for instance, the density or temperature or pressure at the point.

Then f$(x, y, z) = C$, a constant, would imply a relation connecting x, y, z, and would be the equation of a surface; for instance, a surface of constant density.

If V denotes the volume contained in a closed surface S, then

$$V = \iiint dxdydz,$$

the integration including all the infinitesimal elements of volume $dxdydz$ contained in the surface S.

Suppose the density ρ within the surface S to be variable and a given function of x, y, z; then if M denotes the mass contained by the surface S, and $\bar{x}$, $\bar{y}$, ⁻ the co-ordinates of the centre of mass,

$$M = \iiint \rho dxdydz,$$

$$M\bar{x} = \iiint x\rho dxdydz, \quad M\bar{y} = \iiint y\rho dxdydz, \quad M\bar{z} = \iiint z\rho dxdydz.$$

With the same proof as before (§§ 112, 113), denoting f(x, y, z) for brevity by f,

$$\frac{df}{dt} = \frac{\partial f}{\partial x}\frac{dx}{dt} + \frac{\partial f}{\partial y}\frac{dy}{dt} + \frac{\partial f}{\partial z}\frac{dz}{dt};$$

and, expanded in powers of h, k, l,

$$f(x+h, y+k, z+l) = f(x, y, z) + h\frac{\partial f}{\partial x} + k\frac{\partial f}{\partial y} + l\frac{\partial f}{\partial z} + \dots$$

$$+ \frac{1}{n!}\left(h\frac{\partial}{\partial x} + k\frac{\partial}{\partial y} + l\frac{\partial}{\partial z}\right)^n f(x, y, z)$$

$$+ \frac{1}{(n+1)!}\left(h\frac{\partial}{\partial x} + k\frac{\partial}{\partial y} + l\frac{\partial}{\partial z}\right)^{n+1} f(x+\theta h, y+\theta k, z+\theta l);$$

and these theorems can be generalized for a function of any number of independent variables.

The geometrical interpretation of the independent variables will not hold when they are more than three in number without the introduction of the fiction of space of more than three dimensions, a thing which is inconceivable.

A function $f(t, x, y, z)$ of four independent variables, t, x, y, z, may, however, be interpreted, as in Hydrodynamics, as representing the velocity or density or pressure at the time t at a point in space whose co-ordinates are x, y, z.

A rational integral *homogeneous* algebraical function of the nth degree in m variables is defined to be a function in which the sum of the indices of the variables in each term is constant and equal to n, the indices being positive integers; such a function is denoted in Higher Algebra by

$$(x, y, z, t, \ldots)^n,$$

and is called a *quantic*, in m variables and of the nth degree.

The general expression of the quantic is obtained by expanding

$$(x+y+z+t+\ldots)^n$$

by the Multinomial Theorem, and applying an arbitrary constant multiplier, denoted by a, b, c, ..., to each term.

Thus the binary quantic $(x, y)^n$ represents the expression

$$ax^n + nbx^{n-1}y + \frac{n(n-1)}{2!}cx^{n-2}y^2 + \ldots$$

Sometimes the binomial coefficients are omitted, and the quantic is written

$$ax^n + bx^{n-1}y + cx^{n-2}y^2 + \ldots;$$

and sometimes it is convenient to write the quantic in the form
$$ax^n + nbx^{n-1}y + n(n-1)cx^{n-2}y^2 + \ldots,$$
inserting the *derived* coefficients n, $n(n-1)$, $\ldots$; and sometimes in the form
$$ax^n + \frac{b}{1!}x^{n-1}y + \frac{c}{2!}x^{n-2}y^2 + \ldots,$$
inserting the *exponential* coefficients $\frac{1}{1!}, \frac{1}{2!}, \frac{1}{3!}, \ldots$

The first and fourth form of the quantics will be found closely related, as well as the second and third; any *seminvariant* of the first or third form being a *non-unitary symmetric function* of the fourth or second form respectively.

(American Journal of Mathematics, vol. VI., *Seminvariants and Symmetric Functions*, by Captain P. A. MacMahon, R.A.)

Denoting the general quantic in m variables and of the nth degree by u, then
$$x\frac{\partial u}{\partial x} + y\frac{\partial u}{\partial y} + z\frac{\partial u}{\partial z} + \ldots = nu.$$

This is proved by considering a single term of the quantic $x^p y^q z^r \ldots$, where $p + q + r + \ldots = n$, for which the theorem is at once established; for, denoting this term by v, then
$$x\frac{\partial v}{\partial x} = pv, \ y\frac{\partial v}{\partial y} = qv, \ z\frac{\partial v}{\partial z} = rv, \ \ldots.$$

More generally
$$\left(x\frac{\partial}{\partial x} + y\frac{\partial}{\partial y} + z\frac{\partial}{\partial z} + \ldots\right)^2 u = n(n-1)u,$$
$$\left(x\frac{\partial}{\partial x} + y\frac{\partial}{\partial y} + z\frac{\partial}{\partial z} + \ldots\right)^3 u = n(n-1)(n-2)u,$$
and so on.

These were proved by Lagrange by expanding the quantic $u_1 = (x + hx, y + hy, z + hz, \ldots)^n = (1+h)^n u$

in the form

$$u_1 = h + h\left(x\frac{\partial}{\partial x} + y\frac{\partial}{\partial y} + z\frac{\partial}{\partial z} + \dots\right)u$$
$$+ \frac{h^2}{2!}\left(x\frac{\partial}{\partial x} + y\frac{\partial}{\partial y} + z\frac{\partial}{\partial z} + \dots\right)^2 u$$
$$+ \dots\dots\dots$$

and equating coefficients of like powers of h.

These theorems are called *Euler's Theorems of Homogeneous Functions.*

119. *Green's Theorem.*

Consider a fixed closed surface S, and a function X of three independent variables x, y, z, the co-ordinates of a point in space.

Then, the triple integration extending over the volume enclosed by the surface S,

$$\iiint\frac{\partial X}{\partial x}dxdydz = \iint(-X_1 + X_2 - X_3 + \dots)dydz,$$

where X_1, X_2, X_3, ... are the values of X where a point moving from $-\infty$ to $+\infty$ parallel to the axis of x successively enters and leaves the interior of the surface S.

Denoting by l_1, l_2, l_3,... the cosines of the angles the *outward* drawn normals of the surface S at these points make with the axis of x, then

$$dydz = -l_1 dS_1 = l_2 dS_2 = -l_3 dS_3 = \dots,$$

supposing the infinitesimal prism on the base $dydz$ parallel to the axis of x to intercept the elements of surface dS_1, dS_2, dS_3, ..., on entering and leaving the surface S.

Therefore

$$\iiint\frac{\partial X}{\partial x}dxdydz$$
$$= \iint(l_1 X_1 dS_1 + l_2 X_2 dS_2 + l_3 X_3 dS_3 + \dots) = \iint lXdS,\dots\text{(i.)}$$

the double integration extending over the surface S.

Similarly, if Y, Z, are given functions of x, y, z,

$$\iiint \frac{\partial Y}{\partial y} dx dy dz = \iint m Y dS \dots\dots\dots\dots (\text{ii.})$$

$$\iiint \frac{\partial Z}{\partial z} dx dy dz = \iint n Z dS \dots\dots\dots\dots (\text{iii.})$$

where m, n denote the cosines of the angles the *outward* drawn normals of the surface S make with the axes of y, z respectively.

Therefore, adding

$$\iiint \left(\frac{\partial X}{\partial x} + \frac{\partial Y}{\partial y} + \frac{\partial Z}{\partial z} \right) dx dy dz = \iint (lX + mY + nZ) dS. (\text{iv.}).$$

Now suppose U and U' two given functions of x, y, z; then from equation (i.), integrating by parts,

$$\iiint \frac{\partial U}{\partial x} \frac{\partial U'}{\partial x} dx dy dz$$

$$= \iint l U' \frac{\partial U}{\partial x} dS - \iiint U' \frac{\partial^2 U}{\partial x^2} dx dy dz \dots\dots\dots (\text{v.});$$

and therefore

$$\iiint \left(\frac{\partial U}{\partial x} \frac{\partial U'}{\partial x} + \frac{\partial U}{\partial y} \frac{\partial U'}{\partial y} + \frac{\partial U}{\partial z} \frac{\partial U'}{\partial z} \right) dx dy dz$$

$$= \iint U' \left(l \frac{\partial U}{\partial x} + m \frac{\partial U}{\partial y} + n \frac{\partial U}{\partial z} \right) dS$$

$$- \iiint U' \left(\frac{\partial^2 U}{\partial x^2} + \frac{\partial^2 U}{\partial y^2} + \frac{\partial^2 U}{\partial z^2} \right) dx dy dz$$

and therefore by symmetry

$$= \iint U \left(l \frac{\partial U'}{\partial x} + m \frac{\partial U'}{\partial y} + n \frac{\partial U'}{\partial z} \right) dS$$

$$- \iiint U \left(\frac{\partial^2 U'}{\partial x^2} + \frac{\partial^2 U'}{\partial y^2} + \frac{\partial^2 U'}{\partial z^2} \right) dx dy dz \dots\dots (\text{vi.}).$$

Now
$$l\frac{\partial U}{\partial x} + m\frac{\partial U}{\partial y} + n\frac{\partial U}{\partial z} = \frac{\partial U}{\partial \nu},$$
$$l\frac{\partial U'}{\partial x} + m\frac{\partial U'}{\partial y} + n\frac{\partial U'}{\partial z} = \frac{\partial U'}{\partial \nu},$$

where $\frac{\partial U}{\partial \nu}$, $\frac{\partial U'}{\partial \nu}$ represent the rates of variation of U, U' in the direction of the *outward* drawn normal to the surface.

Then, representing the operator $\frac{\partial^2}{\partial x^2} + \frac{\partial^2}{\partial y^2} + \frac{\partial^2}{\partial z^2}$ by $-\nabla^2$ (Maxwell, *Electricity*, chap. I.), the preceding theorem can be written

$$\iiint \left(\frac{\partial U}{\partial x}\frac{\partial U'}{\partial x} + \frac{\partial U}{\partial y}\frac{\partial U'}{\partial y} + \frac{\partial U}{\partial z}\frac{\partial U'}{\partial z} \right) dxdydz,$$

$$= \iint U'\frac{\partial U}{\partial \nu}\,dS + \iiint U'\nabla^2 U\,dxdydz,$$

$$= \iint U\frac{\partial U'}{\partial \nu}\,dS + \iiint U\nabla^2 U'\,dxdydz\ldots\ldots\ldots(\text{vii.});$$

and this is called *Green's Theorem*, a theorem of great use in the mathematical theories of electricity and magnetism.

(*An Essay on the application of Mathematical Analysis to the Theories of Electricity and Magnetism*, by G. Green; edited by N. M. Ferrers.)

The theorem may be given in the more general form

$$\iiint a^2\left(\frac{\partial U}{\partial x}\frac{\partial U'}{\partial x} + \frac{\partial U}{\partial y}\frac{\partial U'}{\partial y} + \frac{\partial U}{\partial z}\frac{\partial U'}{\partial z} \right) dxdydz$$

$$= \iint a^2 U'\frac{\partial U}{\partial \nu}\,dS + \iiint U'\nabla^2(a^2 U)\,dxdydz,$$

$$= \iint a^2 U\frac{\partial U'}{\partial \nu}\,dS + \iiint U\nabla^2(a^2 U')\,dxdydz,$$

where a may be constant, or any given function of x, y, z.

(Thomson and Tait's *Natural Philosophy*, vol. I, appendix A),

CHAPTER VI.

CURVES IN GENERAL.

120. *Equation of the chord, tangent, asymptote, and normal of a curve in polar co-ordinates.*

The implicit relation, $f(r, \theta) = 0$, connecting the polar co-ordinates r and θ of all points on a curve is called the *polar equation* of the curve.

Instead of r it is convenient to use u, the reciprocal of r, and then if u is given as an explicit function of θ by the equation $u = f\theta$, it is required to determine the equation of a chord, tangent, etc.

To find the chord which passes through two points on the curve whose vectorial angles are $a+\beta$ and $a-\beta$, assume that the equation of the chord is of the form
$$u = A \cos(\theta - a) + B \sin(\theta - a),$$
where A and B are the constants to be determined.

That this is the equation of a straight line can be seen by changing to the *Cartesian* co-ordinates x and y, by writing it in the form
$$1 = Ar \cos(\theta - a) + Br \sin(\theta - a)$$
$$= A(x \cos a + y \sin a) + B(y \cos a - x \sin a),$$
since
$$x = r \cos\theta, \quad y = r \sin\theta.$$

Now, when $\theta = a + \beta$, $u = f(a + \beta)$, so that
$$f(a + \beta) = A \cos\beta + B \sin\beta;$$
and when $\theta = a - \beta$, $u = f(a - \beta)$, so that
$$f(a - \beta) = A \cos\beta - B \sin\beta.$$

From these two equations
$$A = \frac{f(a+\beta)+f(a-\beta)}{2 \cos\beta}, \quad B = \frac{f(a+\beta)-f(a-\beta)}{2 \sin\beta};$$

so that the equation of the chord is
$$u = \frac{f(a+\beta)+f(a-\beta)}{2 \cos\beta} \cos(\theta-a) + \frac{f(a+\beta)-f(a-\beta)}{2 \sin\beta} \sin(\theta-a) \ldots \text{(i.)}$$

Now, suppose β to become ultimately zero; the chord then becomes the tangent at the point given by $\theta = a$; and since

$$\mathrm{lt}\frac{f(a+\beta)+f(a-\beta)}{2 \cos\beta} = fa, \quad \mathrm{lt}\frac{f(a+\beta)-f(a-\beta)}{2 \sin\beta} = f'a,$$

therefore the equation of the tangent at $\theta = a$ is

$$u = fa \cos(\theta - a) + f'a \sin(\theta - a) \ldots \ldots \ldots \ldots \text{(ii.)}$$

and $OP = \dfrac{1}{fa}$, $OT = -\dfrac{1}{f'a}$, if $xOP = a$ (fig. 5).

Suppose $fa = 0$, but $f'a$ is finite; the point of contact is then at an infinite distance from the origin, but the tangent remains at a finite distance; the tangent is then called an *asymptote* of the curve, and its equation is

$$u = f'a \sin(\theta - a) \ldots \ldots \ldots \ldots \ldots \text{(iii.)}$$

The equation

$$u = fa \cos(\theta - a) - \frac{(fa)^2}{f'a} \sin(\theta - a) \ldots \ldots \ldots \ldots \text{(iv.)}$$

will represent the normal at the point $\theta = a$; for it is at right angles to the line represented by equation (ii.), and when $\theta = a$, $u = fa$.

If the equation of the curve is given in the form $r = F\theta$, the equation of the normal at the point $\theta = a$ will be

$$\frac{1}{r} = \frac{\cos(\theta - a)}{Fa} + \frac{\sin(\theta - a)}{F'a} \dots\dots\dots\dots\dots(v.),$$

and $OP = Fa$, $OG = F'a$, if $xOP = a$ (fig. 5).

Denoting by p the length of the perpendicular OY from the origin O on the tangent at P, then (§ 17)

$$p = r\sin\phi, \text{ where } \tan\phi = \frac{rd\theta}{dr};$$

and

$$\frac{1}{p^2} = \frac{1}{r^2}\text{cosec}^2\phi = \frac{1}{r^2}\cot^2\phi + \frac{1}{r^2}$$

$$= \frac{1}{r^4}\frac{dr^2}{d\theta^2} + \frac{1}{r^2} = \frac{du^2}{d\theta^2} + u^2.$$

We have proved that in a central field of force (§ 87)

$$P = h^2 u^2\left(\frac{d^2 u}{d\theta^2} + u\right),$$

so that the orbit is a straight line when $P = 0$ and then $\frac{d^2 u}{d\theta^2} + u = 0$; but the orbit is *concave* to the origin when P, and therefore $\frac{d^2 u}{d\theta^2} + u$ is positive, and the orbit is *convex* to the origin when P, and therefore $\frac{d^2 u}{d\theta^2} + u$ is negative.

At a *point of inflexion*, where the curve changes from *concavity* to *convexity* or *vice versa*, $\frac{d^2 u}{d\theta^2} + u$ vanishes and changes sign.

Definition.—A curve is said to be *concave* with respect to a point or line when it lies on the same side of its tangent as the point or line, *convex* with respect to the point or line when it lies on the opposite side; and at a *point of inflexion* the curve crosses the tangent.

Examples.—Find the asymptotes of

(i) $r = a \sec\theta$, (ii) $r = b \operatorname{cosec}\theta$, (iii) $r = a \sec\theta + b \operatorname{cosec}\theta$,

(iv) $r = a \tan\theta$, (v) $r = a \operatorname{cosec}2\theta$, (vi) $r = a \sec 2\theta$,

(vii) $r^2 = a^2 \sec 2\theta$, (viii) $r \sin\theta = a \cos^2\theta$,

(ix) $r \cos\theta = a \cos 2\theta$, (x) $r = a + b \operatorname{cosec}\theta$, (xi) $r\theta = a$,

(xii) $r = \dfrac{a\theta^2}{\theta^2 - a^2}$; and trace the curves.

121. *Pedal Curves.*

The locus of Y, the foot of the perpendicular on the tangent of a curve drawn from the origin O, is called the *pedal* of the curve with respect to O, and O is called the *pole* of the pedal.

Thus the pedal of an ellipse or hyperbola with respect to a focus is the auxiliary circle, and the pedal of a parabola with respect to the focus is the tangent at the vertex of the parabola.

Denoting OY by p, and the angle xOY by ω, then the relation connecting p and ω is the polar equation of the pedal of the curve, with p and ω instead of r and θ as polar co-ordinates.

If OYP is a rigid right angle, of which OY passes through O, and YP touches the curve, then I, the foot of the perpendicular from O on the normal at P, is the *centre of instantaneous rotation* of the right angle; so that IY is the normal of the locus of Y.

Since IY is a diameter of the circle described on OP as diameter, it follows that the envelope of circles described on the variable vector OP as diameter is the pedal with respect to O, the locus of Y.

This can easily be generalized for the case of a rigid angle PYP' touching two fixed curves, at P and P'; the

Q

centre of instantaneous rotation will be at I, the point of
intersection of the normals at P and P', and therefore
IY is the normal of the locus of Y; and since the circle
circumscribing the triangle PYP' has the same normal
at Y, it follows that the locus of Y is the envelope of
these circles.

Returning to the pedal of the curve (fig. 43), then
since the angle OYI is equal to the angle OPI in the same

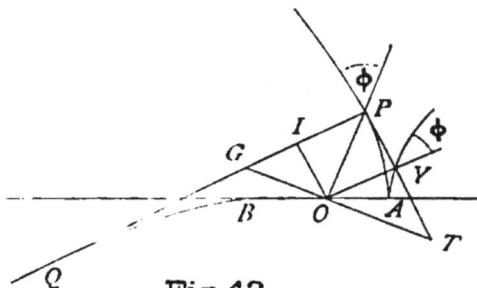

Fig 43

segment, it follows that the pedal curve cuts OY at the
same angle as the curve cuts OP, or the pedal and the curve
have the same angle ϕ (§ 17) at corresponding points.

Therefore
$$\cot \phi = \frac{YP}{OY} = \frac{dp}{p\,d\omega},$$

so that
$$YP = \frac{dp}{d\omega}.$$

The polar co-ordinates of I are therefore $\frac{dp}{d\omega}$ and
$\omega + \frac{1}{2}\pi$; and since Q, the centre of curvature at P, is the
point of contact of the tangent PI of the evolute BQ at
Q, therefore the locus of I is the pedal of the evolute,
and $IQ = \frac{d^2p}{d\omega^2}$; so that the radius of curvature at P

$$\rho = p + \frac{d^2p}{d\omega^2}.$$

These relations can also be proved from the consideration of the relative motion of P and Y; for the component velocities of P in the directions of the normal IP and the tangent YP are 0 and $\dfrac{ds}{dt}$, and the component velocities of Y in the same directions are $\dfrac{dp}{dt}$ and $\dfrac{p\,d\omega}{dt}$ (§ 18); therefore

$$0 = \frac{dp}{dt} - YP\frac{d\omega}{dt},$$

$$\frac{ds}{dt} = \frac{p\,d\omega}{dt} + \frac{d}{dt}(YP).$$

The first equation gives $YP = \dfrac{dp}{d\omega}$, and the second equation gives

$$\rho = \frac{ds}{d\psi} = \frac{ds}{d\omega} = p + \frac{d}{d\omega}(YP) = p + \frac{d^2p}{d\omega^2},$$

since $\qquad\qquad \psi = \omega + \tfrac{1}{2}\pi.$

Also $\qquad\qquad \rho = \dfrac{ds}{d\omega} = \dfrac{ds}{dr}\dfrac{dr}{dp}\dfrac{dp}{d\omega}$

$$= \sec\phi \frac{dr}{dp}YP = \frac{r\,dr}{dp},$$

and the chord of the circle of curvature through the origin is therefore

$$2\frac{p\,dr}{dp}.$$

Now $\qquad\qquad \dfrac{1}{p^2} = \dfrac{1}{r^4}\dfrac{dr^2}{d\theta^2} + \dfrac{1}{r^2} \ (§ 120),$

and therefore, differentiating with respect to θ,

$$-\frac{2}{p^3}\frac{dp}{d\theta} = -\frac{4}{r^5}\frac{dr^3}{d\theta^3} + \frac{2}{r^4}\frac{dr}{d\theta}\frac{d^2r}{d\theta^2} - \frac{2}{r^3}\frac{dr}{d\theta},$$

or $\qquad \left(\dfrac{1}{r^4}\dfrac{dr^2}{d\theta^2} + \dfrac{1}{r^2}\right)^{\frac{3}{2}}\dfrac{dp}{dr} = \dfrac{2}{r^5}\dfrac{dr^2}{d\theta^2} - \dfrac{1}{r^4}\dfrac{d^2r}{d\theta^2} + \dfrac{1}{r^3};$

or
$$\rho = \frac{r\,dr}{dp} = \frac{\left(\dfrac{dr^2}{d\theta^2} + r^2\right)^{\frac{3}{2}}}{2\dfrac{dr^2}{d\theta^2} - r\dfrac{d^2r}{d\theta^2} + r^2}.$$

If the curve passes through the origin, then at O,

$$r = 0, \text{ and } \rho = \tfrac{1}{2}\frac{dr}{d\theta}.$$

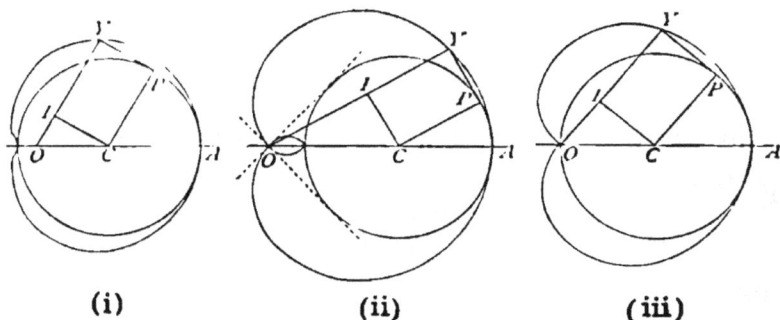

(i)　　　　　　　(ii)　　　　　　　(iii)

Fig. 44

Consider, for example, the pedal of a circle, centre C, with respect to any point O.

Then if $OC = b$, and the radius of the circle is a,

$$p = OY = IY + OI = a + b \cos \omega,$$

the equation connecting the polar co-ordinates p and ω of the locus of Y, and this curve is called a *limaçon* (§ 62).

If $b < a$, O is inside the circle, and the pedal consists of a single oval curve (fig. 44, i.).

If $b > a$, O is outside the circle, and the pedal consists of two loops, one inside the other, having a double point O, where the tangents are the tangents to the circle from O (fig. 44, ii.)

If $b = a$, O is on the circumference, and $p = a\,(1 + \cos \omega)$, the equation of a *cardioid* (fig. 44, iii.).

122. *Roulettes.*

When a curve, carrying a point P fixed to it, rolls on a straight line (or any given curve) the path traced out by the point P is called the roulette of P with respect to the straight line (or given curve).

Thus, when a circle rolls on a straight line, the roulette of a point on the circumference is a *cycloid* (§ 93), and the roulette of any other point fixed in the plane of the circle is called a *trochoid* (§ 62).

An *involute* of a curve (§ 92) is thus the roulette of a point on a straight line which rolls on the curve.

A remarkable analogy, pointed out by Steiner, exists between the roulette of a point with respect to a straight line and the pedal of the rolling curve with respect to the point as pole.

Steiner's Theorems assert that (i.) the length of the arc of the roulette is equal to the length of the corresponding arc of the pedal ; (ii.) the area bounded by an arc of the roulette, the ordinates at the ends of the arc, and the straight line on which the curve rolls is twice the area bounded by the corresponding arc of the pedal and the vectors from the origin to the ends of the arc.

For, if AP is the roulette of the point P when the curve is rolled on the straight line Ox (fig. 45), and if PM is the perpendicular from P on Ox, the tangent at I to the rolling curve, then relatively to P the locus of M is the pedal of the rolling curve with respect to P ; and therefore relatively to M the locus of P is the same curve ; so that we may suppose the pedal $A'P$ rolled on the roulette AP, so that M is always vertically over P if Ox is horizontal ; and the pedal, if loaded so that the centre of gravity is at M, will rest in neutral equilibrium

on the roulette, provided the friction is sufficient to pre-
vent slipping.

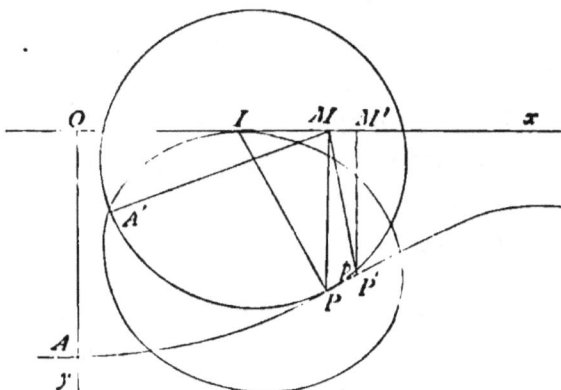

Fig 45

The arc AP of the roulette will then be equal to the
corresponding arc $A'P$ of the pedal, which is Steiner's
first theorem.

For particular examples of this, consult § 62; the
figure here (fig. 45) is drawn for the roulette of the focus
of an ellipse, and the pedal of an ellipse with respect to
a focus, which is a circle.

Also if the pedal is rolled into a consecutive position
so that M comes to M', and the point p of the pedal
comes into contact with the point P' of the roulette,
then the element $MM'P'P$, which is the increment of
area of the roulette, is ultimately double the element
MPp, which is the increment of area of the roulette, or

$$\text{lt } \frac{\text{area } MM'P'P}{\text{area } MPp} = 2 ;$$

and therefore, by integration, the area $OMPA$ of the

roulette is double the area $A'MP$ of the pedal, which is Steiner's second theorem.

This statement of Steiner's second theorem is simpler than that usually given, where the area of the roulette is supposed bounded by the normals at the ends of the arc of the roulette made by a complete revolution of the rolling curve.

Another theorem, also due to Steiner, connects the areas of the different pedals of a closed curve, and therefore of the corresponding roulettes, for different positions of the pole of the pedal.

Let A denote the area of the pedal when the pole is at O the origin, A' the area of the pedal for a different pole O', whose co-ordinates are a, β.

Then
$$A' = \tfrac{1}{2} \int_0^{2\pi} (p - a \cos \omega - \beta \sin \omega)^2 d\omega$$
$$= A - a \int_0^{2\pi} p \cos \omega \, d\omega - \beta \int_0^{2\pi} p \sin \omega \, d\omega$$
$$+ \tfrac{1}{2} \int_0^{2\pi} (a^2 \cos^2\omega + 2a\beta \cos \omega \sin \omega + \beta^2 \sin^2\omega) d\omega.$$

Now $\displaystyle\int_0^{2\pi} \cos^2\omega \, d\omega = \pi,\ \int_0^{2\pi} \cos \omega \sin \omega \, d\omega = 0,\ \int_0^{2\pi} \sin^2\omega \, d\omega = \pi$;

so that
$$A' = A - a \int_0^{2\pi} p \cos \omega \, d\omega - \beta \int_0^{2\pi} p \sin \omega \, d\omega + \tfrac{1}{2}\pi(a^2 + \beta^2).$$

We can make the co-efficients of a and β vanish, so that
$$A' = A + \tfrac{1}{2}\pi(a^2 + \beta^2),$$
by placing the origin O at the centre of mass of the

original curve, supposing the density of the arc proportional to the curvature $\dfrac{d\omega}{ds}$; and then

$$\int_0^{2\pi} x\frac{d\omega}{ds}ds = \int_0^{2\pi} p \cos \omega \, d\omega = 0,$$

$$\int_0^{2\pi} y\frac{d\omega}{ds}ds = \int_0^{2\pi} p \sin \omega \, d\omega = 0.$$

Consider, for example, the pedals of a circle (§ 121); then the area of the circle being πa^2, the area of the limaçon $p = a + b \cos \omega$ will be $\pi a^2 + \frac{1}{2}\pi b^2$; for instance, the area of the cardioid is $\frac{3}{2}\pi a^2$, and therefore, by Steiner's second theorem, the area of the corresponding roulette, the cycloid, is $3\pi a^2$.

Again the pedal of the involute of a circle, when the centre of the circle is the pole, is the spiral of Archimedes, and if the involute rolls on a straight line, the roulette of the centre of the circle is a parabola; this explains the nature of the relation between the parabola and the spiral of Archimedes previously noticed (§ 62).

123. *Epicycloids and Hypocycloids.*

These curves are the roulettes of a point on the circumference of a circle which rolls on the outside or inside of a fixed circle (§ 96).

Let O denote the centre and a the radius of the fixed circle, C the centre and c the radius of the rolling circle; and let I denote the point of contact of the circles; then IP is the normal of the roulette of P, because I is the centre of instantaneous rotation of the rolling circle (fig. 46).

Draw the diameter PCD of the rolling circle, and suppose D originally in contact with the fixed circle at

B, and that P is then at A ; A is called an *apse* or *vertex* of the epicycloid.

If the angle xOI is denoted by θ, the arc $ID=$ arc $IB=a\theta$, so that the angle $ICD=\frac{a}{c}\theta$, and the co-ordinates of P in terms of θ for the epicycloid are

$$x = (a+c)\cos\theta + c\cos\left(1+\frac{a}{c}\right)\theta, \quad y = (a+c)\sin\theta + c\sin\left(1+\frac{a}{c}\right)\theta \; ;$$

and for a hypocycloid, change c into $-c$.

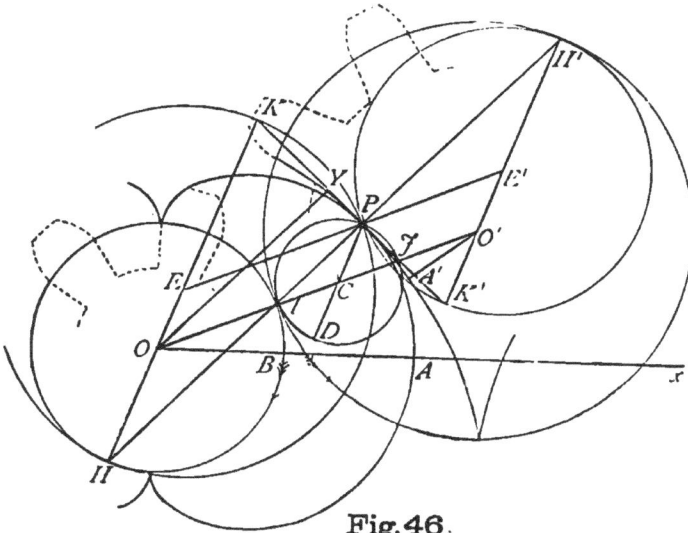

Fig. 46.

Therefore
$$\frac{ds^2}{d\theta^2} = \frac{dx^2}{d\theta^2} + \frac{dy^2}{d\theta^2}$$

$$= (a+c)^2\{\sin\theta + \sin\left(1+\frac{a}{c}\right)\theta\}^2 + (a+c)^2\{\cos\theta + \cos\left(1+\frac{a}{c}\right)\theta\}^2$$

$$= 2(a+c)^2\left(1+\cos\frac{a}{c}\theta\right) = 4(a+c)^2\cos^2\frac{a\theta}{2c} \; ;$$

$$\frac{ds}{d\theta} = 2(a+c)\cos\frac{a\theta}{2c},$$

and, integrating, the arc AP

$$= s = 4\frac{c}{a}(a+c)\sin\frac{a\theta}{2c}.$$

Denoting, as usual, OY the perpendicular from O on the tangent at P by p, and the angle xOY by ω, then

$$p = (a+2c)\cos\frac{a\theta}{2c}, \text{ and } \omega = \theta + \frac{a\theta}{2c},$$

so that

$$p = (a+2c)\cos\frac{a\omega}{a+2c},$$

the polar equation of the pedal of an epicycloid, which is of the form $r = b\cos m\theta$.

Also

$$PY = -\frac{dp}{d\omega} = a\sin\frac{a\omega}{a+2c} = a\sin\frac{a\theta}{2c},$$

so that

$$s = 4\left(\frac{c}{a} + \frac{c^2}{a^2}\right)PY.$$

Again,

$$OP^2 = r^2 = x^2 + y^2$$

$$= (a+c)^2 + 2c(a+c)\cos\frac{a\theta}{c} + c^2,$$

so that

$$r^2 - a^2 = 2c(a+c)\left(1 + \cos\frac{a\theta}{c}\right)$$

$$= 4c(a+c)\cos^2\frac{a\theta}{2c} = 4\frac{c(a+c)}{(a+2c)^2}p^2,$$

the relation connecting p and r in an epicycloid.

The teeth of wheels are usually shaped by epi- and hypo-cycloids; to show how to cut the teeth on two wheels so as to work accurately with each other, produce OI to a fixed point O', and suppose O' the centre of a wheel of radius $O'I = a'$, which revolves without slipping in contact with the wheel of centre O and radius a.

If the circle, centre C and radius c, rolls on the inside of the circle, centre O' and radius a', and describes the hypocycloid $A'P$, then if the vertices A and A' of the

epi- and hypo-cycloid start in contact, the two curves
will roll and slide on each other (fig. 46), so that the
common normal at P passes through I, and therefore the
constant velocity ratio of the wheels is maintained.

Only a small portion of each curve, in the neighbour-
hood of a cusp is made use of to form a tooth; and the
tooth is completed by a portion of an epicycloid on the
circle O', and a portion of a hypocycloid on the circle O,
each described by the rolling of a circle of the same
radius c'.

For instance, if $c' = \frac{1}{2}a$, then the hypocycloid on the
circle O is given by

$$x = (a - c') \cos \theta - c' \cos \left(1 - \frac{a}{c'}\right)\theta = 0,$$

$$y = (a - c') \sin \theta - c' \sin \left(1 - \frac{a}{c'}\right)\theta = a \sin \theta;$$

so that the hypocycloid degenerates into a straight line.

When the radius c of the rolling circle is made infinite,
the epicycloid and hypocycloid both become *involutes* of
the circles O and O', and the teeth of wheels are sometimes
made of this shape; involute teeth have the advantage of
preserving the velocity ratio of the wheels constant for a
variable distance between the centres of the wheels.

Produce PI both ways to meet the circles O and O'
again in H and H', and draw EPE' parallel to OO' to
meet OH, $O'H'$ in E and E'.

Then E and E' are the centres of circles of radii $a + c$
and $a' - c$ which touch each other at P, and the circles O
and O' at H and H', so that the same epicycloid AP and
hypocycloid $A'P$ can be described by the rolling of these
circles on the circles O and O' respectively; this is called
the *double generation of the epicycloid and hypocycloid.*

124. *Inverse Curves.*

If the vector OP of a curve is produced to Q, so that OQ is inversely proportional to OP; or $OP \cdot OQ = c^2$, a constant, then the locus of Q is called an *inverse curve* of the locus of P, with respect to the origin O, or with respect to the circle of centre O and radius c.

Thus, if $u = f\theta$ is the polar equation of the locus of P, then $r = c^2 f\theta$ is the polar equation of the inverse curve, the locus of Q with respect to the origin O.

The inverse of a circle (or sphere) is another circle (or sphere), except when the circle (or sphere) passes through the origin of inversion, when the inverse is a straight line (or a plane).

For if OPQ meets a circle (or sphere) in P and Q, then $OP \cdot OQ = OT^2$, a constant, where OT is a tangent to the circle (or sphere), so that the circle or sphere is its own inverse, with respect to any origin O, and all circles and spheres are similar.

A curve and its inverse cut the vector OPQ at complementary angles; for if P', Q' are corresponding consecutive points on the curve and the inverse,
$$OP' \cdot OQ' = OP \cdot OQ,$$
so that a circle can be described round $PQQ'P'$, and therefore the angles QPP' and $QQ'P'$ are complementary, and PP', QQ' are ultimately the tangents at P and Q.

Otherwise, in the curve described by P,
$$\cot \phi = \frac{dr}{rd\theta} = \frac{d \log r}{d\theta} = -\frac{d \log u}{d\theta} = -\frac{f'\theta}{f\theta};$$
and in the curve described by Q,
$$\cot \phi' = \frac{du}{ud\theta} = \frac{d \log u}{d\theta} = \frac{f'\theta}{f\theta};$$
so that $\cot \phi + \cot \phi' = 0$, or $\phi + \phi' = \pi$.

Consequently if two curves cut at a given angle, the inverse curves cut at the same angle; for instance the inverse of a system of *orthogonal* curves, that is, two sets of curves intersecting at right angles, is another system of orthogonal curves; and the inverse of a system of *oblique trajectories* of orthogonal curves, that is, curves cutting the orthogonal curves at a constant angle, is another system of oblique trajectories of the system of orthogonal inverse curves.

As an exercise, prove that the inverse curves of the dipolar system of circles of § 29 with respect to either pole S or S' are a system of concentric circles, and a system of straight lines through the common centre; and the inverse of the oblique trajectories are equiangular spirals.

125. *Exact Mechanical Parallel Motion.*

When P describes a given curve, the point Q can be made to describe an inverse curve by means of the mechanical invertors of link motion invented by Peaucellier and Hart.

Peaucellier's motion consists of a rhombus LP, PM, MQ, QL formed by four links of equal length, jointed at L, M, P, Q, and two equal links OL, OM, jointed at a fixed point O (fig. 47).

Then, however the link motion is displaced by the motion of P,

$$OP . OQ = OE^2 - EP^2 = OL^2 - LP^2,$$

a constant, so that P and Q describe inverse curves.

For instance, if P is made to describe an arc of a circle passing through O by means of a link CP, jointed at a fixed point C, where $OC = CP$, then Q will move in a straight line perpendicular to *OC*, so that *Q* can be attached to the head of a piston rod, and thus Peaucellier's motion accomplishes with exactness what is only approximately effected by Watt's Parallel Motion.

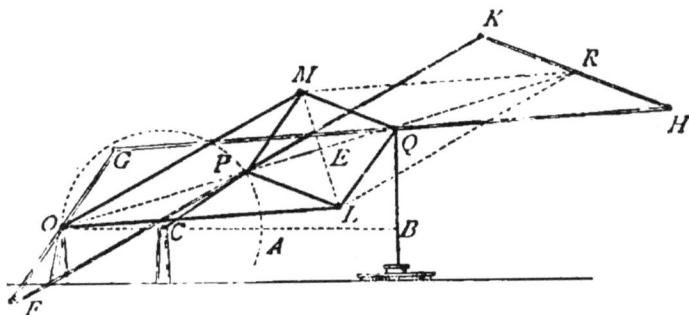

Fig 47

Hart's Parallel Motion accomplishes the same purpose with four bars, while Peaucellier's requires six.

A jointed rhomboid *FGHK* of rods is taken, and the rods are crossed; any fixed point *O* in *FG* is taken and *OPQR* is drawn parallel to *FH* or *GK* to meet *FK* in *P*, *GH* in *Q*, and *HK* in *R*; then *P*, *Q*, and *R* are fixed points in *FK*, *GH*, and *HK* (fig. 47), such that when *FGHK* is opened out into a parallelogram, *OQRP* will be a parallelogram the sides of which are parallel to the diagonals of the parallelogram *FGHK*; and *OP . OQ* is constant, so that *P* and *Q* describe inverse curves if *O* is fixed.

For
$$\frac{OP}{GK} = \frac{FP}{FK}, \quad \frac{OQ}{FH} = \frac{PK}{FK};$$

so that
$$OP . OQ = \frac{FP . PK}{FK^2} FH . GK$$

$$= \frac{FP . PK}{FK^2}(GH^2 - HK^2), \text{ which is constant.}$$

If we draw *FG* parallel to *PM* or *LQ*, *GQH* parallel to *OL*, and *FPK* parallel to *OM*; then *O* is the middle point of *FG*, and *FG* or *HK* is twice the length of a side of the rhombus *LPMQ*, and *FK* and *GH* are twice the length of *OL* or *OM* (fig. 47).

We may join *LR* and *MR* by bars, and then the two rhombuses *LPMQ*, *OLRM* are said to make a complete Peaucellier cell.

When *P* and *Q* are inside the cell, the cell is called *positive* (fig. 47); but when *P* and *Q* are outside, it is called a negative cell (fig. 48).

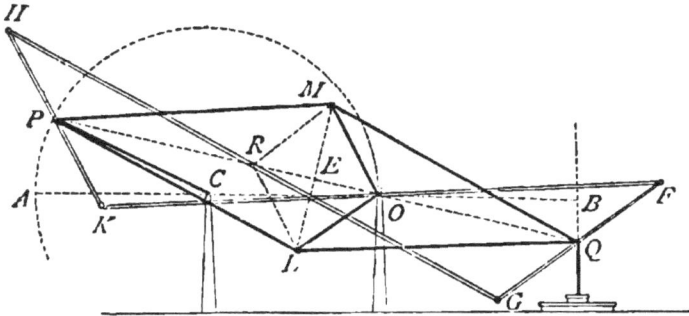

Fig 48

126. *Polar Reciprocals.*

The inverse of the pedal of a curve with respect to the same pole is called the *polar reciprocal of the curve*.

For instance, the pedal of a circle with respect to any point *O* is the limaçon, whose polar equation in *r* and θ is
$$r = a + b \cos \theta,$$
and the polar reciprocal of a circle is therefore the inverse of a limaçon, and its equation is
$$r = \frac{c^2}{a + b \cos \theta},$$
the polar equation of a conic, a focus at the origin.

If $b<a$, the conic is an ellipse, the inverse of an oval limaçon (fig. 44, i.), and the origin O is inside the circle; if $b>a$, the conic is a hyperbola, the inverse of a nodal limaçon (fig. 44, ii.), and the origin O is outside the circle; if $b=a$, the conic is a parabola, the inverse of a cardioid, and the origin O is on the circle (fig. 44, iii.).

The polar equations of the confocal conics (§ 74), with origin at a focus S, are

$$r=\frac{c \sinh^2\eta}{\cosh \eta+\cos \theta}, \quad r=\frac{c \sin^2\xi}{\cos \xi-\cos \theta};$$

and therefore a system of inverse curves is given by the equations

$$r \sinh^2\eta = c(\cosh \eta+\cos \theta), \quad r \sin^2\xi=c(\cos \xi-\cos \theta),$$

the equations of a system of orthogonal limaçons; and these limaçons can be proved to be the pedals of the system of circles (§ 29), which do not pass through S and S', with respect to S or S'.

The polar reciprocals of these circles with respect to S or S' are therefore a system of confocal conics.

The polar reciprocals of the system of circles through S and S' will be a system of parabolas.

A large class of important curves are denoted by the equation

$$r^n = a^n\cos n\theta,$$

where n is some number; for instance the curve is (i.) when $n=1$, a circle; (ii.) $n=-1$, a straight line; (iii.) $n=-2$, a rectangular hyperbola; (iv.) $n=2$, a lemniscate; (v.) $n=\frac{1}{2}$, a cardioid; (vi.) $n=-\frac{1}{2}$, a parabola.

Differentiating logarithmically,

$$\cot\phi = \frac{d\log r}{d\theta} = -\tan n\theta,$$

so that

$$\phi = \tfrac{1}{2}\pi + n\theta,$$

and therefore

$$\omega = (1+n)\theta\,;$$

and

$$p = r\cos n\theta = a(\cos n\theta)^{\frac{1}{n}+1}$$

$$= a\left(\cos\frac{n\omega}{n+1}\right)^{\frac{n+1}{n}},$$

or

$$p^m = a^m\cos m\omega,$$

where $m = \dfrac{n}{n+1}$, the polar equation of the pedal, an equation of the same form as that of the original curve, with m for n.

The equation of an inverse curve with respect to the origin will be

$$r^n\cos n\theta = b^n\,;$$

and of a polar reciprocal

$$r'^n\cos m\theta = c^n.$$

If AP is the roulette with respect to a straight line of the origin of the curve $r^n = a^n\cos n\theta$ (fig. 45), then, denoting the angle IPM by ψ, and MP by y,

$$\psi = n\theta = m\omega,$$

and

$$y = p = a(\cos m\omega)^{\frac{1}{m}},$$

so that

$$y^m = a^m\cos m\omega = a^m\cos\psi,$$

the relation connecting y and ψ in the roulette.

Differentiating logarithmically with respect to s,

$$\frac{m}{y}\frac{dy}{ds} = -\tan\psi\frac{d\psi}{ds},$$

and

$$\frac{dy}{ds} = -\sin\psi,\; \frac{ds}{d\psi} = \rho\,;$$

then
$$\rho = \frac{y \sec \psi}{m} = \frac{PI}{m},$$

so that the radius of curvature of the roulette is $\frac{1}{m}$ or

$1 + \frac{1}{n}$ of the length of the normal PI.

For instance, if $n = 1$, the rolling curve is a circle, and the roulette a cycloid, and $\rho = 2PI$, as before (§ 93).

Suppose the roulette AP is the catenary curve assumed under gravity by a chain of variable density σ per unit of length, with Ox horizontal (fig. 45); then if T denotes the tension at P, T_0 at the lowest point A, and W the weight of the chain AP, resolving horizontally and vertically for the equilibrium of the part AP,

$$T \cos \psi = T_0, \quad T \sin \psi = W;$$

so that
$$W = T_0 \tan \psi;$$

and
$$\sigma = \frac{dW}{ds} = T_0 \sec^2 \psi \frac{d\psi}{ds} = \frac{mT_0}{a}(\sec \psi)^{\frac{1}{n}+1},$$

so that the density at A, where $\psi = 0$, is $\sigma_0 = \frac{mT_0}{a}$;

therefore
$$\frac{\sigma}{\sigma_0} = (\sec \psi)^{\frac{1}{m}+1} = (\sec \psi)^{\frac{1}{n}+2} = \left(\frac{T}{T_0}\right)^{\frac{1}{n}+2}.$$

For instance, if $n = -\frac{1}{2}$, σ is constant, and the ordinary catenary is the roulette of the focus of a parabola.

If m is negative, the roulette AP is convex to the axis of x; put $m = -p$, then

$$y^p = a^p \sec \psi, \text{ and } \rho = \frac{y \sec \psi}{p}.$$

· Now suppose that AP is a trajectory described under gravity in a resisting medium, to determine the retardation R due to the resistance.

Resolving normally (§ 90)

$$\frac{v^2}{\rho} = g \cos \psi, \text{ or } v^2 = g\rho \cos \psi = \frac{gy}{p};$$

and resolving tangentially

$$\tfrac{1}{2}\frac{dv^2}{ds} = g \sin \psi - R;$$

so that $\quad R = g \sin \psi - \tfrac{1}{2}\dfrac{dv^2}{ds} = g\left(1 - \dfrac{1}{2p}\right)\sin \psi.$

Denoting the velocity at A by v_0, then $v_0^2 = \dfrac{ga}{p}$,

so that $\quad \sec \psi = \left(\dfrac{y}{a}\right)^p = \left(\dfrac{v}{v_0}\right)^{2p}, \ \sin \psi = \left\{1 - \left(\dfrac{v_0}{v}\right)^{4p}\right\}^{\frac{1}{2}},$

and $\quad R = g\left(1 - \dfrac{1}{2p}\right)\left\{1 - \left(\dfrac{v_0}{v}\right)^{4p}\right\}^{\frac{1}{2}},$

giving the resistance as a function of the velocity.

Examples.—(1) Prove that the equation of the pedal of an ellipse with respect to the centre is

$$p^2 = a^2 \cos^2 \omega + b^2 \sin^2 \omega,$$

and that the polar reciprocal is another ellipse.

(2) Prove that the polar reciprocal of an epi- or hypo-cycloid with respect to the centre is a Cotes's spiral of the form $r \cos m\theta = b$ (§ 87), and that the roulette of the centre with respect to a straight line is an ellipse (*Ex.* 1, p. 98).

(3) Prove that the vibrations in an epi- or hypo-cycloid are isochronous (§ 94) for a repulsion from or attraction to the centre varying as the distance.

127. *Conjugate Functions and Orthogonal and Oblique Trajectories.*

Two quantities α and β are said to be *conjugate functions* of x and y, if $\alpha + i\beta$ is a function of $x + iy$, where $i = \sqrt{(-1)}$. (Maxwell, *Electricity*, Vol. 1, Chap. XII.)

Let
$$\alpha + i\beta = f(x + iy) = fz,$$

suppose, where
$$z = x + iy.$$

Then
$$\frac{\partial \alpha}{\partial x} + i\frac{\partial \beta}{\partial x} = f'z,$$

$$\frac{\partial \alpha}{\partial y} + i\frac{\partial \beta}{\partial y} = if'z;$$

so that
$$\frac{\partial \alpha}{\partial y} + i\frac{\partial \beta}{\partial y} = i\frac{\partial \alpha}{\partial x} - \frac{\partial \beta}{\partial x},$$

and therefore,
$$\frac{\partial \alpha}{\partial x} = \frac{\partial \beta}{\partial y}, \quad \frac{\partial \beta}{\partial x} = -\frac{\partial \alpha}{\partial y};$$

and consequently,
$$\frac{\partial^2 \alpha}{\partial x^2} + \frac{\partial^2 \alpha}{\partial y^2} = 0, \quad \frac{\partial^2 \beta}{\partial x^2} + \frac{\partial^2 \beta}{\partial y^2} = 0.$$

Also the Jacobian
$$\frac{\partial \alpha}{\partial x}\frac{\partial \beta}{\partial y} - \frac{\partial \alpha}{\partial y}\frac{\partial \beta}{\partial x} = \left(\frac{\partial \alpha}{\partial x}\right)^2 + \left(\frac{\partial \beta}{\partial x}\right)^2 = \left(\frac{\partial \alpha}{\partial y}\right)^2 + \left(\frac{\partial \beta}{\partial y}\right)^2$$
$$= f'(x + iy)f'(x - iy) = R^2,$$

suppose.

Conversely, if $x + iy = F(\alpha + i\beta)$

then
$$\frac{\partial x}{\partial \alpha} = \frac{\partial y}{\partial \beta}, \quad \frac{\partial x}{\partial \beta} = -\frac{\partial y}{\partial \alpha};$$

and
$$\left(\frac{\partial x}{\partial \alpha}\right)^2 + \left(\frac{\partial y}{\partial \alpha}\right)^2 = \left(\frac{\partial x}{\partial \beta}\right)^2 + \left(\frac{\partial y}{\partial \beta}\right)^2 = \frac{1}{R^2};$$

so that if x and y are rectangular co-ordinates, and if s, denotes the arc of a curve whose equation is $\beta=$ const., and s_2 the arc of a curve $a=$ const.,

$$\frac{ds_1}{da}=\frac{ds_2}{d\beta}=\frac{1}{R},$$

and the curves intersect at right angles, thus forming a system of *orthogonal* curves.

Two cases of this kind have been discussed in § 29.

In an *oblique* trajectory, cutting the orthogonal curves a and β at constant angles $\frac{1}{2}\pi+\gamma$ and γ,

$$ds_1\cos\gamma+ds_2\sin\gamma=0,$$

or $$da\cos\gamma+d\beta\sin\gamma=0,$$

and therefore $a\cos\gamma+\beta\sin\gamma=$ const.,

is the equation of an oblique trajectory.

As an exercise, prove that the oblique trajectories of $r^n=a^n\cos n\theta$ are given by $r^n=b^n\cos(n\theta-\gamma)$, and determine the orthogonal and oblique trajectories obtained by putting $a+i\beta=(x+iy)^n,\log(x+iy),\exp(x+iy),\cos(x+iy),$ $\tan(x+iy),\cos^{-1}(x+iy),\tan^{-1}(x+iy)$.

127. *Rational Algebraical Curves in Cartesian Coordinates.*

A number of such curves have already been introduced previously, which presented no difficulty in tracing from their equations; a slight sketch will now be given of a systematic method of treatment, but for a complete account the reader is referred to the treatises on *Curve Tracing* by Frost and Woolsey Johnson.

Given the equation of a curve in the rational algebraical form of the implicit relation

$$f(x,\ y)=0,$$

first arrange the terms in groups of binary quantics of

descending order n, $n-1$, ..., 3, 2, 1, 0, denoted by u_n, u_{n-1}, ..., u_3, u_2, u_1, u_0; so that the equation of the curve becomes

$$u_n + u_{n-1} + u_{n-2} + \ldots + u_3 + u_2 + u_1 + u_0 = 0 ;$$

then n is called the *degree* of the curve.

If u_0 does not vanish, the curve does not pass through the origin; but by changing the origin to a point on the curve we can make u_0 vanish, and then $u_1 = 0$ is the equation of the tangent at the origin.

Also $u_2 + u_1 = 0$ is the equation of a conic section, *osculating* the given curve at the origin.

If u_1 also vanishes, then $u_2 = 0$ represents two straight lines, real or imaginary, through the origin, which are the tangents at the origin ; if real, the origin is a *double point*, and the curve crosses itself; if imaginary, the origin is a *conjugate point*, that is, an isolated point the co-ordinates of which satisfy the equation of the curve.

If u_2 is a perfect square, the tangents are coincident, and the origin is in general a *cusp* (§ 96).

If u_2 also vanishes, then $u_3 = 0$ denotes the tangents at the origin, so that if u_3 has three real linear factors, the origin is a *triple point ;* and so on.

Generally to find the *multiple points* of a curve, that is the points where the curve crosses itself, consider the first derived equation

$$\frac{\partial f}{\partial x} + \frac{\partial f}{\partial y}\frac{dy}{dx} = 0 ;$$

this gives in general a determinate value of $\frac{dy}{dx}$, except when

$$\frac{\partial f}{\partial x} = 0, \text{ and } \frac{\partial f}{\partial y} = 0,$$

when the value of $\frac{dy}{dx}$ becomes *indeterminate*, and the second, or third, ... derived equation must be employed to determine $\frac{dy}{dx}$; so that to determine the multiple points of a curve we must find the values of x and y from the equations $\qquad \frac{\partial f}{\partial x} = 0$, and $\frac{\partial f}{\partial y} = 0$, which also satisfy the equation of the curve

$$f(x, y) = 0;$$

and then having determined these points, a change of origin to such a point will indicate the nature of the point by inspection.

Examples.—Determine the tangents at the origin of the following curves—

1. $x^2(x+y) - a^2(x-y) = 0$.
2. $x^3 + y^3 - 3axy = 0$.
3. $x^2y^2 - a^3(x+y) = 0$.
4. $x^2y^2 - a^2(x^2 - y^2) = 0$.
5. $x^4 + 3ax^2y + 2axy^2 - ay^3 = 0$.
6. $x^4 + y^4 + 6ax^2y - 8ay^3 = 0$.

Secondly, to determine the nature of the curve at an infinite distance from the origin, consider the geometrical interpretation of the equation

$$u_n = 0;$$

which represents n straight lines, real or imaginary, through the origin.

The real straight lines will approximate to the nature of the curve at an infinite distance, and will therefore be parallel to the rectilinear *asymptotes*, if asymptotes exist.

The definition of an *asymptote* has already been given (§ 120); it is a tangent to the curve of which the point of contact is at infinity, the tangent itself remaining at a finite distance from the origin; or it is a straight line to which the curve continually approaches and ultimately at an infinite distance becomes indefinitely near.

The equation of an asymptote will therefore be of the form
$$y = mx + n,$$
when $y = mx$ is the equation of one of the straight lines represented by $u_n = 0$.

The problem of finding an asymptote is then, from the implicit relation $f(x, y) = 0$, to expand y by *reversion of series* in descending powers of x in the form
$$y = mx + n + \frac{p}{x} + \frac{q}{x^2} + \ldots \ldots$$
Substituting this value of y in terms of x in the equation $f(x, y) = 0$, and treating the resulting equation as an identical equation, and equating to zero the coefficients of x^n, x^{n-1}, x^{n-2}, $\ldots \ldots$, sufficient equations are obtained to determine $m, n, p, q, \ldots \ldots$

Then m determines the *direction* and n the *position* of the asymptote, and p or q determines the side of the asymptote on which the curve lies; for this reason it is generally useful to expand y in descending powers of x as far as three terms.

If u_n has a factor x, then to determine the corresponding asymptote we must expand x in descending powers of y in the form
$$x = n' + \frac{p'}{y} + \frac{q'}{y^2} + \ldots \ldots;$$
or we may in general put
$$x = m'y + n' + \frac{p'}{y} + \frac{q'}{y^2} + \ldots \ldots,$$
and determine $m', n', p', q' \ldots \ldots$, as before.

When it is possible to obtain y explicitly in terms of x, or x in terms of y, from the implicit relation $f(x, y) = 0$, the asymptotes are then determined by expanding in descending powers by the Binomial Theorem ; thus if

$$x^3 + y^3 = a^3,$$

then $\quad y = \sqrt[3]{(a^3 - x^3)} = -x\left(1 - \dfrac{a^3}{x^3}\right)^{\frac{1}{3}} = -x + \dfrac{a^3}{3x^2}\ldots\ldots;$

or $\quad x = \sqrt[3]{(a^3 - y^3)} = -y\left(1 - \dfrac{a^3}{y^3}\right)^{\frac{1}{3}} = -y + \dfrac{a^3}{3y^2}\ldots\ldots$

Also, if $x = a$ makes $y = \infty$, or $y = b$ makes $x = \infty$, then $x - a = 0$ is an asymptote, and also $y - b = 0$.

For example, if the equation is

$$\frac{a^2}{x^2} + \frac{b^2}{y^2} = 1,$$

then $\quad y^2 = \dfrac{b^2 x^2}{x^2 - a^2},$ and $x^2 = \dfrac{a^2 y^2}{y^2 - b^2},$

so that $x \pm a = 0$ and $y \pm b = 0$ are asymptotes.

The preceding considerations are in general sufficient for tracing a curve whose equation is given, but considerations of symmetry are also useful; thus, if only *even* powers of x appear in the equation, the curve is symmetrical right and left of the axis of y; if only *even* powers of y appear, the curve is symmetrical above and below the axis of x.

Examples.—Determine the asymptotes of the following curves, and trace the curves.

1. $\dfrac{x^2}{a^2} - \dfrac{y^2}{b^2} = 1.$

2. $x^2 y + xy^2 = a^3.$

3. $y^3 = x^2(2a - x).$

4. $x^3 - xy^2 + ay^2 = 0.$

5. $x^3 + y^3 - 3axy = 0.$

6. $x^2 y^2 = a^2(x^2 + y^2)$
   or $a^2(x^2 - y^2).$

7. $x^4 - y^4 - a^2 xy = 0.$

8. $x^5 + y^5 - 5ax^2 y^2 = 0.$

Trace also the curves of the preceding set of examples.

APPENDIX.

NOTE A.

It is sometimes simpler to use the definition

$$\frac{df x}{dx} = \mathrm{lt}\,\frac{f(x+h) - f(x-h)}{2h},$$

which is equivalent to the ordinary definition (§ 1); for instance, in this way

$$\frac{dx^2}{dx} = \mathrm{lt}\,\frac{(x+h)^2 - (x-h)^2}{2h} = 2x \text{ (§ 3).}$$

$$\frac{d \sin x}{dx} = \mathrm{lt}\,\frac{\sin(x+h) - \sin(x-h)}{2h}$$

$$= \mathrm{lt}\,\cos x\,\frac{\sin h}{h} = \cos x \text{ (§ 4).}$$

$$\frac{d \cos x}{dx} = \mathrm{lt}\,\frac{\cos(x+h) - \cos(x-h)}{2h}$$

$$= -\sin x\,\mathrm{lt}\,\frac{\sin h}{h} = -\sin x \text{ (§ 6).}$$

Again, in this way (§ 34)

$$\frac{d\,uv}{dx} = \mathrm{lt}\,\frac{(u+\Delta u)(v+\Delta v) - (u-\Delta u)(v-\Delta v)}{2\Delta x}$$

$$= \mathrm{lt}\left(\frac{\Delta u}{\Delta x}v + u\frac{\Delta v}{\Delta x}\right) = \frac{du}{dx}v + u\frac{dv}{dx}.$$

NOTE B.

In the proof of $\dfrac{dx^{n-1}}{dx} = nx^{n-1}$ (§ 3) the Binomial Theorem and its convergency have been assumed; but the proof can be given without this assumption.

(i.) Suppose n a positive integer, and denote $x+h$ by x_1; then

$$\frac{dx^n}{dx} = \operatorname{lt} \frac{x_1{}^n - x^n}{x_1 - x}$$
$$= \operatorname{lt}(x_1{}^{n-1} + xx_1{}^{n-2} + \ldots + x^{n-2}x_1 + x^{n-1}) = nx^{n-1}.$$

(ii.) Suppose n a positive fraction $\dfrac{p}{q}$, and put $x = z^q$, $x_1 = z_1{}^q$; then

$$\frac{dx^n}{dx} = \operatorname{lt}\frac{x_1{}^{\frac{p}{q}} - x^{\frac{p}{q}}}{x_1 - x}$$
$$= \operatorname{lt}\frac{z_1{}^p - z^p}{z_1{}^q - z^q}$$
$$= \operatorname{lt}\frac{z_1{}^{p-1} + zz_1{}^{p-2} + \ldots + z^{p-2}z_1 + z^{p-1}}{z_1{}^{q-1} + zz_1{}^{q-2} + \ldots + z^{q-2}z_1 + z^{q-1}}$$
$$= \frac{pz^{p-1}}{qz^{q-1}} = \frac{p}{q}z^{p-q} = nx^{n-1}.$$

(iii.) Suppose n a negative number $-m$; then

$$\frac{dx^n}{dx} = \operatorname{lt}\frac{x_1{}^{-m} - x^{-m}}{x_1 - x}$$
$$= -\operatorname{lt}\frac{x_1{}^m - x^m}{(x_1 - x)x_1{}^m x^m} = -\frac{mx^{m-1}}{x^{2m}} = -mx^{-m-1} = nx^{n-1};$$

so that $\dfrac{dx^n}{dx} = nx^{n-1}$, universally.

INDEX.

(The numbers refer to the pages.)

GLASGOW : ROBERT MACLEHOSE, PRINTER TO THE UNIVERSITY.

RECENT
MATHEMATICAL PUBLICATIONS.

A TREATISE ON THE CALCULUS OF VARIATIONS. Arranged with the purpose of Introducing, as well as Illustrating, its Principles to the Reader by means of Problems, and designed to present in all important particulars a complete view of the present state of the Science. By LEWIS BUFFETT CARLL, A.M. Demy 8vo. 21s.

CONSTRUCTIVE GEOMETRY OF PLANE CURVES. With numerous Examples. By T. H. EAGLES, M.A., Instructor in Geometrical Drawing, and Lecturer in Architecture at the Royal Indian Engineering College, Cooper's Hill. Crown 8vo. 12s.

A TREATISE ON DIFFERENTIAL EQUATIONS. By ANDREW RUSSEL FORSYTH, M.A., Fellow and Assistant Tutor of Trinity College, Cambridge. Demy 8vo. 14s.

AN ELEMENTARY TREATISE ON THE INTEGRAL CALCULUS, founded on the Method of Rates or Fluxions. By WILLIAM WOOLSEY JOHNSON, Professor of Mathematics at the United States Naval Academy, Annopolis, Maryland. Demy 8vo. 8s.

CURVE TRACING IN CARTESIAN CO-ORDINATES. By WILLIAM WOOLSEY JOHNSON, Professor of Mathematics at the United States Naval Academy, Annopolis, Maryland. Crown 8vo. 4s. 6d.

AN ELEMENTARY TREATISE ON THE DIFFERENTIAL CALCULUS, founded on the Method of Rates or Fluxions. By JOHN MINOT RICE, Professor of Mathematics in the United States Navy, and WILLIAM WOOLSEY JOHNSON, Professor of Mathematics at the United States Naval Academy. Third Edition, Revised and Corrected. Demy 8vo. 16s. Abridged Edition. 8s.

MACMILLAN AND CO., LONDON.

A TREATISE ON THE DYNAMICS OF A SYSTEM OF RIGID BODIES. By E. J. ROUTH, D.Sc., LL.D., F.R.S. With numerous Examples. Fourth Edition, revised and enlarged. 8vo. In Two Parts. Part I. Elementary, 14s. Part II. Advanced, 14s.

DIFFERENTIAL AND INTEGRAL CALCULUS, with Applications. By ALFRED GEORGE GREENHILL, M.A., Professor of Mathematics to the Senior Class of Artillery Officers, Woolwich; Examiner in Mathematics in the University of London. Crown 8vo. 7s. 6d.

A TEXT BOOK ON THE METHOD OF LEAST SQUARES. By MANSFIELD MERRIMAN, Professor of Civil Engineering at Lehigh University, Member of the American Philosophical Society, etc. 8vo. 8s. 6d.

APPLIED MECHANICS. An Elementary General Introduction to the Theory of Structures and Machines. By JAMES H. COTTERILL, F.R.S., Professor of Applied Mechanics in the Royal Naval College, Greenwich. Medium 8vo. 18s.

Nature says that the volume "bears on every page evidence that its author has not only studied and become intimately acquainted with his subject, but that he possesses the rare faculty of having learned by experience in teaching the best way of presenting a subject so as to diminish its difficulties and make rough places smooth for the footsteps of the beginner."

PHYSICAL ARITHMETIC. By A. MACFARLANE, M.A., D.Sc., F.R.S.E., Examiner in Mathematics to the University of Edinburgh. Crown 8vo. 7s. 6d.

This treatise may be described as a treatise on Applied Arithmetic, the applications being chiefly in Physical Science. Knowledge of the elements of pure arithmetic is assumed, but the more advanced methods are explained when their application happens to occur. The subject is treated under the following heads: (1) Financial; (2) Geometrical; (3) Kinematical; (4) Dynamical; (5) Thermal; (6) Electrical; (7) Acoustical; (8) Optical; and (9) Chemical.

MACMILLAN AND CO., LONDON.

A COLLECTION OF EXAMPLES ON HEAT AND ELECTRICITY. By H. H. TURNER, B.A., Fellow of Trinity College, Cambridge. Crown 8vo. 2s. 6d.

CONIC SECTIONS. By CHARLES SMITH, M.A., Fellow and Tutor of Sidney Sussex College, Cambridge. Second Edition. Crown 8vo. 7s. 6d.

AN ELEMENTARY TREATISE ON SOLID GEOMETRY. By CHARLES SMITH, M.A., Fellow and Tutor of Sidney Sussex College, Cambridge. Crown 8vo. 9s. 6d.

ELEMENTARY ALGEBRA. By CHARLES SMITH, M.A., Fellow and Tutor of Sidney Sussex College, Cambridge. Crown 8vo. [*Shortly.*

ELEMENTARY ALGEBRA FOR SCHOOLS. A New Algebra for Schools. By H. S. HALL, B.A., formerly Scholar of Christ's College, Cambridge, Master of the Military and Engineering Side, Clifton College, and S. R. KNIGHT, B.A., formerly Scholar of Trinity College, Cambridge, late Assistant Master of Marlborough College. Globe 8vo. · 3s. 6d. With Answers, 4s. 6d.

> *Nature* says, "This is, in our opinion, the best *Elementary* algebra for school use. . . . We confidently recommend it to mathematical teachers, who we feel sure will find it the best book of its kind for teaching purposes. Many subjects of interest are also treated of, and a vast collection of (3,500) examples will fur- · nish ample exercise for the boys, and save the teacher the trouble of concocting illustrations of the best methods."

WEEKLY PROBLEM PAPERS. With Notes. Intended for the use of Students preparing for Mathematical Scholarships, and for the Junior Members of the Universities who are reading for Mathematical Honours. By Rev. JOHN J. MILNE, Second Master of Heversham Grammar School. 18mo. 3s. 6d.

Questions on the following subjects only have been admitted : Algebra, Arithmetic, Euclid, Trigonometry, Geometrical Conics, the Elementary Parts of Analytical Conics, Statics, and in a few of the latter papers Dynamics.

MACMILLAN AND CO., LONDON.

SOLUTIONS OF WEEKLY PROBLEM PAPERS. By the Rev. JOHN J. MILNE, M.A. Crown 8vo. 10s. 6d.

DIFFERENTIAL CALCULUS FOR BEGINNERS. With a Selection of Easy Examples. By ALEXANDER KNOX, B.A., Cantab. 18mo. 3s. 6d.

> *Nature* says, "This little book deserves a hearty welcome from those who are engaged in leading forward students to the higher mathematics. . . . Presenting a carefully-selected set of illustrations of infinitesimals, limits, and differential co-efficients, which a student may profitably work through before entering upon the usual formal treatises on the calculus. We know of no work in English comparable with the present since De Morgan's *Elementary Illustrations of the Differential and Integral Calculus.*"

ELEMENTARY TRIGONOMETRY. By Rev. J. B. LOCK, M.A., Senior Fellow, Assistant Tutor and Lecturer in Mathematics, of Gonville and Caius College, Cambridge; late Assistant Master at Eton. Globe 8vo. 4s. 6d.

HIGHER TRIGONOMETRY. By the same Author. Globe 8vo. 4s. 6d. Both Parts complete in One Volume. Globe 8vo. 7s. 6d.

ARITHMETIC FOR SCHOOLS. By Rev. J. B. LOCK, M.A. Globe 8vo.
[*In the Press.*

WOOLWICH MATHEMATICAL PAPERS, for Admission into the Royal Military College, Woolwich, 1880—1884 inclusive. Crown 8vo. 3s. 6d.

SOLID GEOMETRY AND CONIC SECTIONS. With Appendices on Transversals and Harmonic Division. For the Use of Schools. By Rev. J. M. WILSON, M.A., Head Master of Clifton College. New Edition. Extra fcap. 8vo. 3s. 6d.

MACMILLAN AND CO., LONDON.